Joseph Parrish Thompson

Man in Genesis and in Geology

The Biblical Account of Man's Creation Tested by Scientific Theories of his Origin

and Antiquity

Joseph Parrish Thompson

Man in Genesis and in Geology
The Biblical Account of Man's Creation Tested by Scientific Theories of his Origin and Antiquity

ISBN/EAN: 9783337415600

Printed in Europe, USA, Canada, Australia, Japan

Cover: Foto ©berggeist007 / pixelio.de

More available books at **www.hansebooks.com**

IN GENESIS AND IN GEOLOGY:

Or, the Biblical Account

OF

MAN'S CREATION,

TESTED BY

Scientific Theories

OF HIS ORIGIN AND ANTIQUITY.

By JOSEPH P. THOMPSON, D.D., LL.D.

NEW YORK:
SAMUEL R. WELLS, PUBLISHER,
No. 389 Broadway.
1870.

TO

JAMES D. DANA, LL.D.,

Professor in Yale College,

AS A TRIBUTE TO HIS

EMINENT ATTAINMENTS IN SCIENCE, AND IN GRATEFUL
RECOGNITION OF HIS SERVICES

IN ILLUSTRATING THE

HARMONY OF TRUTH

IN THE WORKS AND THE WORD OF GOD,

This Volume

IS INSCRIBED BY HIS FRIEND,

THE AUTHOR.

PREFACE.

The question *How to adjust the facts of Science to the Bible?* assumes not only that the Bible is a book of divine authority, but that its authority reaches over the world of physical phenomena with which Science is directly concerned, so that no fact declared by Science can be accepted as true if it conflicts with any statement of the Bible. The question *How to adjust the Bible to the facts of Science?* assumes that the Bible is constantly on trial, in respect of its truth and its divine authority; and that in any case of apparent conflict, the facts of Science must take precedence of the declarations of the Bible. Hence, on the one hand, the cry of infidelity is raised against men of Science, and on the other the Bible is set aside, at least in all that relates to the primeval history of the world and Man, as a book of crude and antiquated traditions. Either of these modes of viewing the relations of the Bible and Science is incomplete and illogical. The true method of physical Science keeps within its own province of the observation and induction of facts, and will not trespass upon the ground of Biblical criticism and interpretation. A sound Theology looks upon Nature as the handiwork of God, and while it accepts a supernatural Revelation upon evidence peculiar to itself, it accepts also every established fact of the physical universe as equally of divine origin and authority. Hence the devout inquirer after truth will be bent,—not upon devising some compromise between Science and the Bible, as presumably at variance,—but upon ascertaining the exact facts of Nature, as a portion of God's testimony concerning Himself, and the precise meaning of the Bible according to legitimate principles of interpretation. When each class of declarations is fairly brought out by its own methods, if there is a seeming discrepancy, neither will be set aside as of inferior authority, but either some error of observation, induction, or interpretation will be suspected; or while both forms of testimony are accredited, the decision of the case will be held in abeyance, until a more advanced knowledge shall reconcile them from some higher plane, where the harmonies of all Science, physical and metaphysical, and of all Revelation, the secondary and the supernatural, shall interblend without confusion or mistake. It is from this last point of view that this book has been written. It is neither a book of Science nor of Theology, but it aims to present the latest results of Science touching the origin and antiquity of Man, and his place in this mundane system, side by side with the account of his creation and functions in the book of Genesis, as interpreted by the critical tests of modern philology; and to suggest certain principles of adjustment between the record of Nature and the record of the Bible, without violence to the spirit of either.

The matter of the volume was originally given in a series of Sunday-evening lectures, largely extemporaneous in form, and purposely popular, almost colloquial, in style. At the instance of the publisher, these have been prepared for the press from the reports of a competent and careful phonographer. No attempt has been made to elaborate them for scientific readers, though a few notes of reference to authorities and of ancillary topics have been added. The fourth lecture, on MAN'S DOMINION OVER NATURE, is somewhat more labored than the rest, having been delivered substantially to the Phi Beta Kappa Society in Harvard College, in 1865. The then recent death of Mr. EDWARD EVERETT naturally suggested the tribute to his memory as a typical Man.

If this little book shall do anything to diffuse sound views of the interpretation of the Bible in its allusions to the phenomena of Nature, and to strengthen the conviction that in Nature and the Bible alike one living and eternal God is declared the creator and lord of all, and Man His image as a spiritual power above Nature, the author will be fully recompensed for the risk of entering the lists as a disputant in an untried field.

Having in view always the popular reader, the author has cited foreign authorities from English translations, wherever these exist, or has clothed their thoughts in English dress. Among American authors he acknowledges his special indebtedness to Professor JAMES D. DANA, of Yale College, and Professor ARNOLD GUYOT, of the College of New Jersey—men whom Science recognizes among her wisest Interpreters, and Revelation among her ablest Defenders.

NEW YORK, *September*, 1869.

CONTENTS.

LECTURE I.

THE OUTLINE OF CREATION IN GENESIS, 9—Moses the Author of Genesis, 11—Origin of the Universe, 13—Biblical Idea of Creation, 15—Meaning of the Word Day, 17—Outline of Creation in Genesis, 19—Ancient Cosmogonies, 21—Cosmogony of the Veda, 23—The Genesis of Things Revealed by God, 25—Outline of Creation in Genesis, 27.

LECTURE II.

THE CREATION OF MAN, 28—Harmony of Genesis and Geology, 29—Man the Image of God, 33—Man the Head of the Creation, 35.

LECTURE III.

THE ORIGIN OF MAN, 36—Progressive Order not Development, 37—Successive Creations of Species, 39—Progress by Spiritual Power, 41—No Transitional Forms, 43—The Characteristics of Man, 45—Man Distinguished by the Brain, 47—The Dignity of Man, 49.

LECTURE IV.

MAN'S DOMINION OVER NATURE, 51—Man not a Product of Nature, 53—Serial Progression not Evolution, 55—No Links of Development, 57—Man the Conqueror of Nature, 59—Man the only Inventor, 61—Christianity a Civilizing Power, 63—Laws of Nature are God's Volitions, 67—Instinct not a Reasoning Intelligence, 69—Consciousness a Ground of Certainty, 71—The Nobility of Virtue, 73—Edward Everett, a Typical Man, 75—Professor Owen on Species, 77—Owen and Darwin Compared, 78—No Spontaneous Generation, 80—The Supernatural the Highest Science, 82.

CONTENTS.

LECTURE V.

THE ANTIQUITY OF MAN, 84—True Science belongs to Theology, 85—Date of the Pyramids, 86—Pile-Habitations of the Swiss Lakes, 88—Mounds and Peat in Germany, 90—Caution in Framing or Receiving Theories, 93—Did the Human Race begin in Barbarism? 95—No Universal Stone Age, 96—Usher's Chronology too Short, 99—Antiquity of the Negro Race, 101—Man at the Close of the Glacial Period, 103—Adam a Typical Man, 105—Man the Latest and Highest Work, 107—Some Recent Works on Man, 109.

LECTURE VI.

THE SABBATH MADE FOR MAN, 111—The Glory of the Heavenly Host, 112—Rest, the Suspension of Creative Energy, 114—The Origin of the Week, 116—The Reason of the Sabbath Perpetual, 119—The Sabbath a Sanitary Provision, 121—The Sabbath for Spiritual Life, 123.

LECTURE VII.

WOMAN AND THE FAMILY, 125—The Origin of Language, 126—Marriage a Primeval Institution, 128—Sex Fundamental in Human Society, 130—The Family Founded in Love, 132—Mutual Adaptations of the Sexes, 134—The Social Compact a Fiction, 136—Woman more than a *Femmehomme*, 138—Woman's Sex her Spiritual Prerogative, 140—Woman Disqualified by Nature, 142—Woman Rules by Spiritual Prerogatives, 144—How to Elevate the Poor, 146—The Biblical Views of God, 148.

MAN: IN GENESIS AND IN GEOLOGY.

LECTURE I.

The Outline of Creation in Genesis.

1. In the beginning God created the heaven and the earth.
2. And the earth was without form, and void; and darkness *was* upon the face of the deep: and the Spirit of God moved upon the face of the waters.
3. And God said, Let there be light: and there was light.
4. And God saw the light, that *it was* good: and God divided the light from the darkness.
5. And God called the light Day, and the darkness he called Night: and the evening and the morning were the first day.
6. And God said, Let there be a firmament in the midst of the waters; and let it divide the waters from the waters.
7. And God made the firmament, and divided the waters which *were* under the firmament from the waters which *were* above the firmament: and it was so.
8. And God called the firmament Heaven: and the evening and the morning were the second day.
9. And God said, Let the waters under the heaven be gathered together unto one place, and let the dry *land* appear: and it was so.
10. And God called the dry *land* Earth; and the gathering together of the waters called he Seas: and God saw that *it was* good.
11. And God said, Let the earth bring forth grass, the herb yielding seed, *and* the fruit tree yielding fruit after his kind, whose seed *is* in itself, upon the earth: and it was so.
12. And the earth brought forth grass, *and* herb yielding seed after his kind, and the tree yielding fruit, whose seed *was* in itself, after his kind: and God saw that *it was* good.
13. And the evening and the morning were the third day.
14. And God said, Let there be lights in the firmament of the heaven, to divide the day from the night; and let them be for signs, and for seasons, and for days, and years:
15. And let them be for lights in the firmament of the heaven to give light upon the earth: and it was so.
16. And God made two great lights; the greater light to rule the day, and the lesser light to rule the night: *he made* the stars also.
17. And God set them in the firmament of the heaven to give light upon the earth,

1*

18. And to rule over the day and over the night, and to divide the light from the darkness: and God saw that *it was* good.

19. And the evening and the morning were the fourth day.

20. And God said, Let the waters bring forth abundantly the moving creature that hath life, and fowl *that* may fly above the earth in the open firmament of heaven.

21. And God created great whales, and every living creature that moveth, which the waters brought forth abundantly after their kind, and every winged fowl after his kind: and God saw that *it was* good.

22. And God blessed them, saying, Be fruitful, and multiply, and fill the waters in the seas, and let fowl multiply in the earth.

23. And the evening and the morning were the fifth day.

24. And God said, Let the earth bring forth the living creature after his kind, cattle, and creeping thing, and beast of the earth after his kind: and it was so.

25. And God made the beast of the earth after his kind, and cattle after their kind, and everything that creepeth upon the earth after his kind: and God saw that *it was* good.

We have here one of the oldest written documents in the world, perhaps the oldest written account of the creation.* There are monuments and even literary remains of the Egyptians and the Chinese that claim a higher antiquity; but these are, for the most part, dry details of names and numbers, with no consecutive narrative of events, or they are myths, traditions, and religious rituals in the form of poetry. This document is professedly a history, given in historical form, and it concerns the origin of Mankind.

It is commonly ascribed to Moses as its author, either as composer or compiler. Modern criticism has attempted to displace Moses from this traditional position, and to substitute for him historians of later date, perhaps of the time of Solomon, or even as late as the time of the Captivity. It is not essential to the authenticity of the record that we should be able to fix definitely upon its author; but the same proofs of genuineness exist in this case as in respect to the works of Herodotus, Homer, and other writers of great antiquity. The pre-

---

* For the art of writing among the Hebrews consult Hengstenberg on "The Authenticity of the Pentateuch," vol. i., p. 344; Dr. W. Smith, "The Book of Moses," vol. i. p. 13; Ewald, "History of Israel," vol. i., p. 48; Delitzsch, *Commentar über die Genesis*," p. 20; Smith's "Dictionary of the Bible," Art. "Writing;" Bunsen, "Egypt's Place in History," vol. i., p. 306; also vol. iii., p. 394, for the origin of writing among the Chinese; Rawlinson's Herodotus, ii., p. 305.

sumption in any such case is, that the author to whom a work has been ascribed by long and almost unbroken tradition was the real author; and internal evidences may go to substantiate the antiquity and authenticity of the work, even if the name of the author be left in dispute. The art of writing was certainly known in the time of Moses. Monuments of Egypt which antedate the Exodus, exhibit abundant specimens of writing on stone, and some papyrus rolls still extant probably date from a higher antiquity than this book of Genesis. So far, therefore, as the style of the book as a written composition is concerned, it may have been produced at the time of Moses. Ewald, the keenest of critics and the most learned of skeptics concerning the authorship of the Pentateuch as a whole, does not hesitate to ascribe to Moses the tables of the Law, and the substantial groundwork of the system that bears his name; while to account for the production as a whole he invents theories which task credulity much more severely than the notion that it was the single compilation of Moses himself. The grand simplicity of style, and the rough poetic strength in some passages of these early narratives, point to the remoteness of their origin. This Ewald also concedes—regarding such passages as the primitive materials around which the composition, as a whole, clusters.*

Some critics regard the book of Genesis as a mixed composition, made up of different documents. This notion is based upon diversities of style and a marked difference in the name

---

* Ewald, "History of Israel," says: "The two stone tables of the Law are, according to all evidences and arguments, to be ascribed to Moses" (vol. i., p. 48); and again: "Among the long and numerous laws referred to Sinai in the extant narratives, many, particularly among those relating to details, may have sprung up, or at all events have assumed their present form, in the next following age. But those essential truths and social arrangements which constitute the motive power of the whole history must certainly have been there promulgated and firmly ordained." (P. 530.) Ewald assigns the blessing of Jacob and the song of Lamech to a high antiquity—the latter "actually pre-Mosaic." (Vol. i., pp. 70 and 267.)

of God, as used in separate sections. These are now commonly distinguished as the Elohistic and the Jehovistic. Such diversities do exist, and give a plausible foundation for the theory of separate authorship.* The composer of Genesis, as we possess it, may have worked up materials already extant in the form either of oral traditions or of written documents, and in so doing he may not have departed from the original structure of the documents before him, nor attempted to harmonize their phraseology and contents except in a general way; but, notwithstanding these apparent diversities, a law of unity pervades the whole book in its leading conception and its evident purpose, and this points to an essential unity of authorship. The great thought of the book is to exhibit God in connection with the religious and providential history of mankind, and the evident purpose of the early portion is to lay a foundation in history for that Theocracy which was finally developed in Israel. Keeping this in mind, we find it less difficult to trace harmony in the book as a whole than when we confine ourselves to the niceties of literary criticism. Indeed, the moral unity seems quite to overbear the apparent literary diversities, and the latter are scarcely greater than one single author might have indulged in while combining several antecedent documents or traditions into one comprehensive whole. But the critical niceties of this question

---

* "Admitting this distinction, we may still doubt whether it has not been carried to an unwarrantable extent. It reduces the Old Scriptures not only to fragments, but to fragments of fragments, in most ill-assorted and jumbled confusion. Surely no other book was ever so composed or so compiled. In the same portion, presenting every appearance of narrative unity, some critics find the strangest juxtapositions of passages from different authors, and written at different times, according as the one name or the other is found in it. There are the most sudden transitions even in small paragraphs having not only a logical but a grammatical connection. One verse, and even one clause of a verse, is written by the Elohist, and another immediately following by the Jehovist, with nothing besides this difference of names to mark any difference in purpose or in authorship. Calling it a compilation will not help the absurdity, for no other compilation was ever made in this way."—*Dr. Tayler Lewis in Lange's* "*Genesis*," p. 107.

can not be popularized, and we assume as sufficient for our purpose the substantial oneness of Genesis as a work of Moses. It is more important to trace the internal evidences of the truth of the narrative and its divine origin.

The subject of this first chapter is the origin of the existing order of things—the earth and its inhabitants, with the visible surrounding heavens. This is one of the profoundest subjects of human thought. It has occupied the speculations of the greatest philosophers of ancient times, and the investigations and theories of modern science ; but neither philosophy nor science has yet accurately determined the origin of the universe. The method of Genesis is the reverse of physical science. The latter, by induction, seeks after laws, principles, and causes; but Genesis begins with the great First Cause. Science leads us back step by step to the necessity of an original cause; Genesis sets that cause before us directly in the declaration, "In the beginning God created the heaven and the earth." If the account given of the creation in this chapter is true, it must have proceeded from God. There was no human observer to record it, and the facts are beyond human discovery even in the present advanced stage of science. It is impossible to believe that, in the age of the world when this book was composed, and among the people to whom it was first given, the human mind should have been capable of originating such a description of the universe. It was communicated by illumination from God to man. The truth of this will appear if we look at it somewhat in detail.

"*In the Beginning!*" This describes a vague period before the present condition of things had an existence, before the heaven and earth, as they now are, began to be. There is here no limitation of time, and therefore the expansion of astronomical and geological eons, cycle upon cycle, finds here the most ample scope. There was time enough in that "Beginning" for the evolution of the entire solar system from a single nebulous mass—supposing that to have been the condition in which matter was first produced.

"*God created the heaven and the earth!*" Did the writer mean to describe the universe at large and the origin of matter? or simply our globe and its visible firmament, as established or constituted in its existing order? This can not be determined from the word *bara*, which has the same ambiguity as the English word *create;* but inasmuch as the succeeding verses are occupied with the plastic process in detail, by which crude chaotic matter was reduced to form and order, we may infer that by the act of creation in the first verse was intended the origination of matter, the first beginning of that from which the worlds were shaped. This is the meaning put upon it by the writer of the epistle to the Hebrews:— "the worlds were framed by the word of God, so that things which are seen were not made of things which do appear."[*] The objection that the metaphysical notion of creation *ex nihilo* is foreign to the Scriptures, has little weight, since the Hebrew writer, impressed with the eternal self-existence and the absolute personality of God, was declaring a fact, without reference to a philosophical mode of conceiving that fact.

As used in the Bible, the word *bara* sometimes signifies the bringing into existence a *new* thing—as, for instance, the creation of Matter, of Life, and of Man—and sometimes the constituting or establishing in order that which had already been brought into existence as to its germs or essence—in the sense to *cut, carve,* or *shape;* but in either case the principle is the same—a personal God giving existence, form, and order to matter by his own power and will. Applied to the acts of the Almighty, *bara* always denotes the giving existence to

---

[*] Hebrews xi. 3.

something *new*,* either in substance or in form, and the bringing into being by divine power is the leading idea in creation. This verse represents God as the primary cause of the whole material creation which comes under the observation of our senses, and which is comprehensively described as "the heaven and the earth."†

At first we have a picture of *chaos:*—matter in a crude, formless condition, shrouded in darkness. The first act of the divine will, represented as "the Spirit brooding upon the waters," is the evolution of light. A beautiful experiment has been invented to illustrate the possible formation of the world from a gaseous condition, according to the nebular theory. In a globe of water and alcohol, mixed in a nicely proportioned density, is deposited a diminutive ball of oil, which, by its relative specific gravity, adjusts itself to the center of the fluid mass. A certain motion imparted to this by a wire from without gives it the shape of our globe flattened at the poles; another motion will throw off the moon, or, if you please, the four moons of Jupiter; again, Saturn and its rings may be produced by another rotary movement; and finally, the whole mass broken up into globules representing the planetary system as it swims in space.

Our knowledge of the prodigious force of gases, and of the effects of motion and electricity on a grand scale, may help

---

* An important passage for the meaning *to create out of nothing* is Genesis ii. 3, where, according to Gesenius, we read, "he rested from all his work which God created in making: *i. e.,* which he made in creating something new; see also Jer. xxxi. 22; whence it is apparent that *bara* implies the creation of something new, not before existing." This view is ably advocated by Dr. Barrows in the *Bibliotheca Sacra* for 1856. p. 743; by Kalisch, "Commentary on Genesis;" by Delitzsch, "*Commentar über die Genesis*," and others; but Dr. Taylor Lewis, in Lange's "Genesis" (p. 127). maintains that the word denotes, not the primal origination, but formations, dispositions, of matter. Yet he adds, "this is creation; it is the divine supernatural making of something new, and which did not exist before."

† Keeping in mind the Hebrew conception of one, eternal, almighty, self-existent God, the natural interpretation of *bara* would appear to be the bringing something out of nothing, although in its strict metaphysical form the doctrine of creation *ex nihilo* can hardly be traced in the early Hebrew Scriptures.

us to understand how, if this Chaos was matter in a rare gaseous state diffused in space, molecular motion, or a chemical change evolving electricity, may have produced the light here described, and then motion, once set in order, might have given shape by degrees to the earth and the heavenly bodies. As to the process, however, all is mere conjecture; Genesis does not describe it,—science can not unfold it.*

Here comes in the term "*Day.*" I suppose it now to be well understood that neither this word itself, nor Biblical usage, nor the context here, requires us to understand by a Day a period of twenty-four hours. The term is first applied to the appearing of light after the darkness of chaos. Chaos was the evening, light the morning. But when did this darkness begin? and how long did the light thus engendered continue? Was this merely a natural day? Why should we attempt to measure this first period by a chronometer which, according to the narrative itself, could not have come into use until the fourth day, when the heavenly bodies became visible from our globe, so as to serve for the measurement of times and seasons?

---

* The nebular hypothesis is thus stated by Prof. Loomis: "Suppose that the matter composing the entire solar system once existed in the condition of a single nebulous mass, extending beyond the orbit of the most remote planet. Suppose that this nebula has a slow rotation upon an axis, and that by radiation it gradually cools, thereby contracting in its dimensions. As it contracts in its dimensions, its velocity of rotation, according to the principles of Mechanics, must necessarily increase, and the centrifugal force thus generated in the exterior portion of the nebula would at length become equal to the attraction of the central mass. This exterior portion would thus become detached, and revolve independently as an immense zone or ring. As the central mass continued to cool and contract in its dimensions, other zones would in the same manner become detached, while the central mass continually decreases in size and increases in density. The zones thus successively detached would generally break up into separate masses, revolving independently about the sun; and if their velocities were slightly unequal, the matter of each zone would ultimately collect in a single planetary but still gaseous mass, having a spheroidal form, and also a motion of rotation about an axis. As each of these planetary masses becomes still farther cooled, it would pass through a succession of changes similar to those of the first solar nebula; rings of matter would be formed surrounding the planetary nucleus, and these rings, if they broke up into separate masses, would ultimately form satellites revolving about their primaries. . . . . . This hypothesis must be regarded as possessing considerable probability, since it accounts for a large number of phenomena which hitherto had remained unexplained."—*Treatise on Astronomy*, p. 314.

In the fourth verse of the second chapter we have an example of the use of this word "Day" to cover the whole period of operations included in the seven days of the first chapter: "These are the generations of the heavens and the earth when they were created, *in the day* that the Lord God made the earth and the heavens." Here the whole term of creation is comprehended within one day. Again: we are told that "one day is with the Lord as a thousand years, and a thousand years as one day."* In short, the word is used in the Scriptures to describe an event or period which had a beginning and a completion. Lest any should suppose that this interpretation of the word Day is a modern invention to accommodate the narrative in Genesis to the discoveries of Geology, or to evade the objections of science to this record, let me remind you that Augustine, in the fourth century, by the simple principles of interpretation, called these "ineffable days," describing them as alternate births and pauses in the work of the Almighty—the boundaries of periods in the vast evolution of the worlds.† And such was the earlier Christian interpretation of this narrative. The notion that these were literal days of twenty-four hours seems rather to have sprung up in the middle ages, an offspring of that literalism and realism which in times of ignorance have often perverted the meaning of the Scriptures.

It has been objected to this narrative, that the sun, moon, and stars did not appear until the fourth day, whereas the growth of vegetation requires the action of light, and the light of certain stars requires to travel for ages before reaching an observer on our earth; and therefore there must have been light from the heavenly bodies during the period of vegetable growth described as the third day, and the stars

---

* 2 Peter iii. 8.   † "*De Genesi ad Literam.*"

must have existed for ages before, in order that their light might at this time have become visible. But there is in all this no conflict with the account in Genesis, if we remember that the language of this narrative is popular and not scientific. The description is optical or phenomenal, that is, of things as they would have appeared, or may be imagined to have appeared, to a human observer, could one then have been stationed on the earth. Vegetation of course required light, and the existence of light has already been announced from the first day. There was cosmical light even when the shining of the sun and other heavenly bodies was not apparent. Let us suppose a human observer (though we well know that man could not have existed in that primitive condition of the globe) to have been stationed on the earth during the period of the vegetation which produced the coal deposits—when the globe was wrapped in dense steaming mists. The sun would have been no more visible than through a London fog! If after a long experience of this condition of the earth and its atmosphere the observer had seen these mists rolling away, the atmosphere gradually clearing up, the light beginning to break in from above—his first glimpse of the sun shining in the distant heavens would appear to him as a new creation, and in optical or popular language he would properly describe the sun, the moon, the stars, then first made visible, as *created* upon that day.

How was this language understood by those to whom it was originally addressed? By disregarding that principle of interpretation which seeks the meaning of an author in the familiar conceptions of his own age, and forcing upon his words ideas derived from later discoveries and other modes of thought, great violence has been done to the text and teaching of Moses. "The great majority of readers," says Max Muller, "transfer without hesitation the ideas which they connect with words as used in the nineteenth century to the mind of Moses or his contemporaries, forgetting altogether the distance which divides their language and their thought from the thoughts and language of the wandering tribes of Israel."*

Without going further into details, it is enough to say that a principle of order and of progress runs through the narrative, whose main features correspond wonderfully with the best results of Geology. We must bear in mind that Geology, one of the newest of sciences, has already many times changed its own theories of the order and method of the structure of our globe; but that order which is now generally accepted by the most accomplished geologists—of whom Guyot, Dana, and Agassiz may be taken as types—is substantially as follows:— that the first movement toward the present condition of things was the beginning of activity in matter, as this was already diffused in a chaotic, perhaps a gaseous, state. This activity was attended with the evolution of light. Next, the earth was divided from the fluid that surrounded it, and assumed a condition of solidity. Next, its features began to appear in outline; then vegetable life, characterized in Gen. i. 11 as "having seed in itself," organic matter in distinction from inorganic substances of which the earth was previously composed. Fourth, there came in light from the sun, having reference to higher systems of life, then about to appear upon the globe. Fifth, the lower orders of animals were introduced in a successive series, and finally appeared the mammals— and man, the crown and end of the whole. This outline, sketched by science, is in remarkable correspondence with that given in the first chapter of Genesis; for what we have in Genesis is simply an outline. The writer does not give the processes of creation, but the succession of phenomena, and

---

* "Chips from a German Workshop," I., p. 133.

his object at every step is to exhibit the power of God. Each central thought, each advancing step in the series, is brought out with simplicity and boldness to illustrate the glory of the Creator.

How came the writer of this account by such a doctrine of the origin of things? Here is a phenomenon in literature, in the history of the human mind, that the skeptic must account for. Moses knew nothing of Geology; perhaps he did not even apprehend the full meaning of that which he recorded as a vision of the six days. How came it to pass that, in that far antiquity, he laid down a basis of the creation which is in such wondrous harmony with that which science now reveals? Compare this narrative with the cosmogonies of the leading nations of antiquity. There are certain general points of resemblance which only render more striking and impressive the characteristic features in which this differs from those. For instance, the cosmogony of the Babylonians represents the beginning of things as in darkness and water, where nondescript animals, hideous monsters, half-men and half-beasts, appeared, and after this, a woman—who personates the creative spirit or principle—was split into two parts, and the heaven and the earth produced by the division. Then Belus, the supreme divinity, cut off his own head, and his blood trickling down and mingling with the dust of the earth, produced human creatures having intelligence and spiritual life. According to the Phœnician cosmogony, that which first appeared was an ether or a mist diffused in space. Then arose the wind, the representative of motion, and from this agitation proceeded a spiritual God, from whom again in turn proceeded an egg—which is so common a feature of the cosmogonies of antiquity—the division of which, as in the case of the woman, produced the heavens and the earth. The noise of thunder awakened beings into spiritual life. The Egyptian cosmogony was in general harmony with the Phœnician. Its principal divinity was Ptah, the world-creating power, who shaped the cosmic egg, which again appears here, as in the Phœnician. There followed from Ptah a long succession of gods, with various offices and powers—solar, telluric, psychical—from whom at length proceeded demigods, and from these again heroes, until the link of our common humanity was established.

The bare statement of these systems must convince one that Moses borrowed nothing from them, though he was probably familiar with their common conception of the origin of the universe; and the question remains, How was it that he avoided their errors and extravagances, and gave with such severe simplicity a description of the creation, which, for popular uses, no rhetoric could improve and no science can gainsay? It will not meet this question to bring down the date of the composition of Genesis, as Ewald proposes, to the time of Solomon, for the physical history of the globe as now deciphered by Geology was not comprehended in the wisdom of Solomon, and the record that lay hidden in the rocks was no more suspected then than when Moses wandered in the rocky wilderness of Sinai. Besides, at that period, we find no improvement in the prevalent conception of the origin of the universe; but comparing the narrative in Genesis with the cosmogony of Homer and Hesiod, are still compelled to ask, Whence came that unique, exact, sublime account of the creation contained in this book?

According to Grote, "the mythical world of the Greeks opens with the gods, anterior as well as superior to man; it gradually descends, first to heroes, and next to the human race. Along with the gods are found various monstrous natures, ultra-human and extra-human, who can not with propriety be called gods, but who partake with gods and man in the attributes of free-will, conscious agency, and susceptibility of pleasure and pain—such as the Harpies, the Gorgons, the Sirens, the Sphinx, the Cyclops, the Centaurs, etc." * After violent contests among these gigantic creatures and forces, there arises a stable government of Zeus, the chief among the gods. First appears Chaos, then the broad, firm, flat Earth, with deep and dark Tartarus below, and from these proceed various divinities and creatures, some grand and terrible, some simply monstrous; their relations to each other violate all notions of decency and morality; their wars and slaughters, their gross and abominable crimes, issue in successive creative products upon the earth, which terminate at last in the appearing of man. We can not suffer the mythology of the Greeks to be read in our schools, except in expurgated editions; and although at the original basis of this was much poetic beauty of conception and even a sublime spirituality of thought, the representation in the concrete is so gross and offensive, and the details are so contrary to the known facts of science, that both our moral sense and our intelligence repudiate it as an account of the origin of the world and of man.

In like manner, should we analyze the cosmogonies of all antiquity, we should find in them certain elements of spiritual thought, grand and imposing, an approximation to the truth as the highest religion and philosophy now give it, but intermingled with this much that is puerile, grotesque, absurd or gross—the intervention of the egg, of the tortoise, of the elephant, of a variety of mundane or monstrous creatures and powers in evolving the principles of nature. The defect of all these systems is, that in attempting to describe the process of creation, first, *metaphysically*, they introduce some defec-

---

* "History of Greece," vol. i., chap 1.

tive and even repulsive conception of the Deity and of the spiritual world; and next that, *physically*, they contravene the simplest facts of science. How came it to pass, then, I repeat, that the writer or compiler of this narrative in Genesis, confessedly one of the most ancient cosmogonies of the world, himself familiar with the cosmogony of Egypt, and probably with those of Phœnicia and of other nations farther East, wrote an account which is not only entirely free from the frivolous, absurd, and monstrous representations of parallel cosmogonies, but is in essential accord with the discoveries and developments of modern science? and that throughout he holds the thought steadily to the conception of one supreme, absolute, eternal, spiritual Creator? In its clear and positive conception of God as the creator, this Mosaic cosmogony far surpasses the sublime but mystic hymn of the Veda upon the same theme—one of the earliest relics of Hindu thought and devotion.

> "Nor Aught nor Naught existed; yon bright sky
> Was not, nor heaven's broad woof outstretched above.
> What covered all? what sheltered? what concealed?
> Was it the water's fathomless abyss?
> There was not death—yet was there naught immortal;
> There was no confine betwixt day and night;
> The only One breathed breathless by itself;
> Other than It there nothing since has been.
> Darkness there was, and all at first was veiled
> In gloom profound—an ocean without light.
> The germ that still lay covered in the husk
> Burst forth, one nature, from the fervent heat.
> Then first came love upon it, the new spring
> Of mind—yea, poets in their hearts discerned,
> Pondering, this bond between created things
> And uncreated. Comes this spark from earth
> Piercing and all-pervading, or from heaven?
> Then seeds were sown, and mighty powers arose—
> Nature below, and power and will above.
> Who knows the secret? who proclaimed it here,
> Whence, whence this manifold creation sprang?

> The gods themselves came later into being.
> Who knows from whence this great creation sprang?
> He from whom all this great creation came,
> Whether his will created or was mute,
> The Most High Seer that is in highest heaven,
> He knows it—or perchance even He knows it not." *

This passage, though free from grotesque and absurd combinations of the spiritual and the material, is pantheistic throughout, and while it places the manifoldness of the material creation before the creation of spiritual powers, it hardly concedes to "the One," "the IT" whose breath interpenetrates all existence, a consciousness of the beginning of the creation that somehow proceeded from Itself. Contrast with this the conception of the personal Creator and the description of His work with which Genesis opens. Think how much is asserted in the very first sentence of this book. "It assumes," says Dr. Murphy, "the existence of God, for it is He who in the beginning creates. It assumes His eternity, for He is before all things; and as nothing comes from nothing, He himself must have always been. It implies His omnipotence, for He creates the universe of things. It implies His absolute freedom, for He begins a new course of action. It implies His infinite wisdom; for a *kosmos*, an order of matter and mind, can only come from a being of absolute intelligence. It implies His essential goodness, for the Sole, Eternal, Almighty, All-wise, and All-sufficient Being has no reason, no motive, and capacity for evil. It presumes Him to be beyond all limit of time and place, as He is before all time and place. * * * * * * *

This simple sentence denies atheism; for it assumes the being of God. It denies polytheism, and, among its various forms, the doctrine of two eternal principles, the one good

---

* The Rig-Veda, Book X., Hymn 129; translated in Max Muller's "Chips from a German Workshop," vol. i., p. 76.

and the other evil; for it confesses the one Eternal Creator. It denies materialism, for it asserts the creation of matter. It denies pantheism, for it assumes the existence of God before all things, and apart from them. It denies fatalism, for it involves the freedom of the eternal being." *

Again I call upon the skeptic to answer, Whence came this sublime conception of God, which has never been exceeded by any philosophy since? Whence this wondrously true and accurate outline of the course of creation, in an age of the world when there was no philosophy nor science equal to such conceptions and discoveries—in an age when all the wisdom of the world upon such matters has shown itself to have been utterly and hopelessly at fault? Whence came this account of the creation but from *God* himself, by direct communication to man?

If it be asked how such a communication was made, we can answer only by conjecture. A probable conjecture is, that what here is given in narrative passed before the mind of the original narrator in a series of retrospective visions; that it was a panoramic optical presentation; as in a prophetic vision, future events are made to pass before the mind in a scenic form. As, for instance, the grand series of events described by John in the Apocalypse moved before him in a succession of visions, so this series of phenomena in the course of creation may have been pictorially represented to the mind of the historian in the inverted order of prophecy, and at each shifting of the scene appeared the hand of God!

Moses has not attempted to teach astronomy or geology, nor to anticipate the deductions of any science, physical or metaphysical. But he has here laid down the first fundamental truth in all theology—a personal Creator: "In the

---

* "Commentary on Genesis," i. 1.

beginning God created the heaven and the earth." The existence of God is assumed, yet the universe here contemplated as the work of creative intelligence becomes a convincing argument for the being of God. Can a man walk this earth so manifestly prepared for his abode, enjoy its beauties, appropriate its uses, analyze its mysteries, and not feel that there is a God? Can a man look upon these heavens, measure the distance, the density, the capacity of each star, prescribe the motions of the planets, and summon to light new worlds to explain the aberrations of the old, and not feel that there is a hand divine that binds the sweet influences of Pleiades and looses the bands of Orion, that brings forth Mazzaroth in his season, and guides Arcturus with his sons?

Shall a man look upon himself, and behold how fearfully and wonderfully he is made, and not know that he is God's workmanship? Shall he make a watch, and not perceive that a superior intelligence must have made the delicate organ that keeps time within his own breast? Shall he make a telescope, and not perceive how much higher skill was requisite to make the eye which he so rudely imitates, and without which his telescope would be a worthless tube of tin? Shall he imagine that matter has done for itself what he with all his intelligence and ingenuity can not do with matter? Shall he bring down light from the stars, and not see that it is *God's* light?

Or shall he look within himself? Shall the thinking I, the living soul, which knows that it is not self-existent, that it has not existed from eternity, shall that soul ask itself whence it came, and not feel the spontaneous, glowing response, "I am the offspring of God?" How can a man be an atheist? be an atheist, and yet be a *man?* Can he know himself and not know God? God is seen and felt in all His works, whether man will see Him or no. We have no need to say, "Oh, that I knew where I might find Him!" If we feel after Him, we shall surely find Him, "seeing He is not far from every one of us—for in Him we live, and move, and have our being." "The *fool* hath said in his heart, There is no God."

We wander back in quest of the origin of our race and of the world we inhabit, till we meet this sublime declaration, *In the beginning*, GOD. We traverse the whole field of speculative philosophy, and reach the same sublime result, *In the beginning*, GOD. We roam through the interminable ages and cycles of ages in the eras of geology, and the weary mind comes at length to the same terminus, *In the beginning*, GOD. We take the nebular theory, and melt down the earth to a fluid mass, and evaporate this into the thinnest ether diffused in space, and requiring age upon age of motion to give it solidity and form; we ask whence came the ether? IN THE BEGINNING, GOD. Everywhere it is written, There is a God—a living God, a personal God, a present God. Can there be a higher object of thought than to know such a God? Can there be a higher privilege of love than to know God as a friend?

LECTURE II.

The Creation of Man.

26. And God said, Let us make man in our image, after our likeness: and let them have dominion over the fish of the sea, and over the fowl of the air, and over the cattle, and over all the earth, and over every creeping thing that creepeth upon the earth.

27. So God created man in his *own* image, in the image of God created he him; male and female created he them.

28. And God blessed them, and God said unto them, Be fruitful, and multiply, and replenish the earth, and subdue it: and have dominion over the fish of the sea, and over the fowl of the air, and over every living thing that moveth upon the earth.

29. And God said, Behold, I have given you every herb bearing seed, which *is* upon the face of all the earth, and every tree, in the which *is* the fruit of a tree yielding seed; to you it shall be for meat.

30. And to every beast of the earth, and to every fowl of the air, and to every thing that creepeth upon the earth, wherein *there is* life, *I have given* every green herb for meat: and it was so.

31. And God saw every thing that he had made, and, behold, *it was* very good. And the evening and the morning were the sixth day.

Gen. ii. 7. And the LORD God formed man *of* the dust of the ground, and breathed into his nostrils the breath of life; and man became a living soul.

THESE passages present to us the last stage in the creation, the creation of Man. Before proceeding to this topic, however, we will briefly recapitulate what was said in the previous lecture. It should be a fixed principle of interpretation that Genesis is written in popular and not in scientific language. Had it been written in scientific language it would have defeated its own object as a communication for the benefit of mankind at large. In that early period of the world it would have been as unintelligible as would a discourse upon the magnetic telegraph or the spectrum to the Feejee islanders. Had Moses described the *Brachiopods*, the *Selacians*, the *Ophidians*, the *Saurians—Megalosaur, Palæosaur, Ichthyosaur, Iguanodon*, etc.—the *Palæotherium, Dinotherium, Mastodon*, and so on through the whole nomenclature of modern Geology, his account of the creation would have remained for ages a sealed book, and have passed from the memory of mankind long before the key to its interpretation had been discovered. A revelation in such language would have defeated its own end. The same would have been true of a scientific description of the process of creation. But the account of the creation as actually given is presented optically, as the work might have appeared to an imaginary human observer.

It is equally important to keep in mind that the narrative was given mainly for a moral purpose—to set forth God in human history, and hence there is a grand principle of unity and order in the composition, notwithstanding diversities of phraseology and style. We have seen, also, that there is no contradiction between this narrative of creation and the established facts of science. There have been scientific theories, no doubt, which were contrary to the Mosaic account of creation; and certain interpretations of the book of Genesis have also been contrary to established facts of science; but setting aside merely speculative theories on the one hand, and erroneous interpretations on the other, we find in this narrative, as an outline of the creation, a general harmony with the geological order. The first two days describe chemical action upon inorganic matter; the third day announces the production of vegetative life;—the process of evaporation is still going forward, and the excess of moisture in the atmosphere would, up to this period, have obscured the planetary bodies;—but on the fourth day the astronomical heavens are made visible in their relation to our globe; the fifth and sixth days introduce the successive gradations of animal life that culminate at last in man.

Two or three points in this narrative are worthy of more particular notice than was given in the previous lecture, as illustrating the substantial harmony of Geology with Genesis. In the twentieth verse of the first chapter we read, "And God said, Let the waters bring forth abundantly the moving creature that hath life;" and in the succeeding verse we are told that "God created great whales, and every living creature that moveth, which the waters brought forth abundantly, after their kind." The "whales" were more properly monsters of the reptile species; the term is comprehensive, including fishes, serpents, dragons, crocodiles. Now, Geology has taught us that the earliest animals and plants of the globe were wholly water species. There was a long marine era, followed by an amphibian era, in which reptiles and birds were the dominant animal types. All this accords exactly with the statement in Genesis:—the rocks testify now to swarming myriads in the sea, and again to abundance of "flying things," whether insect, bird, or flying reptile, all of which occur in the era succeeding the marine. Here is a wondrous harmony. Again: we know that vegetation was a necessary prelude to animal life, vegetation being directly and largely the food of animals; and this accords with the statement in Genesis, that the plant kingdom was instituted before the creation of animals.

Two remarkable correspondences between the account in Genesis and the facts of Geology concerning the introduction of light are noted by Professor Dana. Science teaches that light is produced by a disturbed action or combination of molecules. It is a result of molecular change. Matter in an inactive state, without force, would be dark, cold, and dead. The first effect of the mutual action of its molecules would be the production of light. The command, "let light be," was, therefore, the summons to activity in matter, and here Genesis is in exact accordance with the teachings of science. The Spirit of God moved upon or breathed over the vast deep—else an abyss of everlasting night—and light, as the essential phenomenon of matter in action, flashed instantly through space. But, although the sun, moon, and stars must have had their places in the physical universe when the earth was established, for a long period the earth was shrouded in its own vapors and warmed with its own heat, and therefore there was no sun nor moon "for days and seasons." When the sun first broke through the clouds, it was a day of joy to the world, standing as one of the grand epochs of its history.

Now, mere human invention would naturally have placed the sun first in order as the source of light. The idea of the appearance of light on the first day, independently of the shining sun, and of the subsequent unvailing of the sun by dispersing the mists and clouds, is a result of modern scientific research, and so foreign to the natural conceptions of the human mind in the early period of its history, that we must ascribe this marvelously exact statement in the first of Genesis to some higher origin. Thus what, upon the face of it, was a seeming discrepancy in regard to the first appearing of the sun, becomes one of the highest confirmations of the truth of the record.

The absence of all puerility and absurdity from this account was also commented upon, and attention was directed to the principle of order which runs through it in describing the course of creation. This principle itself is scientific, as is also the recognition of the great first cause—the personal God. To sum up all on this point—of the harmony of Geology and Genesis—we may adopt the language of Professor Arnold Guyot: "The first thought that strikes the scientific reader is the evidence of Divinity, not merely in the first verse of the record, and the successive fiats, but in the whole order of creation. There is so much that the most recent readings of science have for the first time explained, that the idea of man as the author becomes utterly incomprehensible. By proving the record true, science pronounces it divine, for who could correctly narrate the secrets of eternity but God himself? Moreover, the order or arrangement is not a possible intellectual conception, although we grant to man the intuition of a God. Man would very naturally have placed the creation of vegetation, one of the two kingdoms of life, after that of the sun, and next to that the other kingdom of life, especially as the sunlight is so essential to growth; and the creation of quadrupeds he would as naturally have referred to the fifth day, leaving a whole day to man, the most glorious of all creations. . . . . The creation consists, according to the record, of two great periods; the *first three* days constitute the *inorganic* history, the *last three* days the *organic* history, of the earth. Each period begins with *light:* the *first* light cosmical, the *second* light to direct days and seasons on the earth. Each period ends in a day of two great works. *On the third day* God *divided the land from the waters,* and He saw it was good. Then followed a work totally different, *the creation of vegetation,* the institution of a kingdom of life. So, *on the sixth day*, God created *quadrupeds*, and pronounced His work good; and as a second and far greater work of the day, totally new in its grandest element, He created Man."*

This act of creation, described in the twenty-sixth verse, opens a new chapter in this marvelous history. It is introduced with a new formula; instead of the phrase "And God said," or "And God made," with which the previous acts of creation were introduced, we now read "Let *us* make man *in our image.*" Moreover, other forms of organic life were made, each "after its kind".— a phrase describing the several species of vegetable and animal life.

---

* Prof. Guyot, as cited with comments by Prof. Dana, in the *Bibliotheca Sacra* for January, 1856.

But the type of man was not found in existing organizations, but in the Creator himself; "Let us make man in our image, *after our likeness.*" What are we to understand by man's resemblance to God as his image? Certainly not a likeness in outward appearance; for although the Scriptures figuratively ascribe to God the members of the human body, hands, eyes, feet, etc., no one imagines that there is any such resemblance of form between the Creator and the creature. Neither was man created, in the proper sense of that word, in the likeness of God in respect to character, for holiness is not properly a subject of physical creation. Besides, we read in the epistle of James concerning man in his fallen condition, that he still retains his original likeness to God:—"The tongue can no man tame; it is an unruly evil, full of deadly poison. Therewith bless we God, even the Father, and therewith curse we men, *which are made after the similitude of God.*"

We must seek this resemblance in man's intellectual constitution, in his spiritual capacities and powers, in his moral faculties, and in that position of dominion in which he was placed to represent the Creator upon the earth. Man is a reasonable, personal soul, and in this respect is the likeness of God. As the Psalmist expresses it: "Thou madest him a little lower than Elohim"—the "angels" as it reads in our version; but the word used by David was properly the name of God himself. The secondary meaning of "angels,"—which is quoted from the Septuagint in the epistle to the Hebrews—was probably chosen by the Greek translators on account of the superstitious reverence of the Jews for the name of God. The literal statement of the Psalmist concerning man is, Thou madest him but little short of the Divine.

The whole physical creation was prepared as a platform for man, as a temple for its priest; and his likeness to God appears in the supremacy with which he was invested over nature, by virtue of his spiritual power. As Tholuck expresses it: "The lion has his tooth, the crocodile his coat of mail, the birds their wings, the fish their fins; but which is man's weapon for attack, which his shield for defense?—the *spirit from God:* therefore all must obey him. The cattle on the pasture, wild beasts roaming the forests, birds flying below the expanse of heaven, fish swimming in the depths of the sea; they all must obey him—man is their lord and king." *

The seventh verse of the second chapter of Genesis points expressly to the dual constitution of man—an *animal* nature formed of the dust of the ground, that is, a physical organization from existing materials, and the *spiritual* nature, the divine in-breathing, by virtue of which man, as a conscious and rational soul, resembles God as a being of intelligence, having power of voluntary action, and invested with dominion over nature.

Such is the characterization of man in the Biblical account of his origin. Geology assigns to man the same position in the order of creation which is given him in the book of Genesis. Upon this point all geologists, however diverse their theories, are perfectly agreed. Man *began* to be. In certain stages of our globe he could not have existed on its surface. For instance, in the carboniferous period, when rank vegetation flourished over the regions of our present coal beds, the atmospheric conditions of temperature and moisture, and the excess of carbonic acid in the air, were conditions impossible to human life. For long, long periods there is no trace of man's existence in the strata of the globe, nor are his remains found among the earlier fossils of organic forms. Man is the highest type of organization upon the

---

* "Commentary on Psalm viii."

surface of the globe. In particular members, and in adaptations to particular ends, other creatures are superior to man; but by his powers of locomotion, of endurance, and of contrivance, man is fitted to subdue all other creatures, and to subjugate and modify the earth. Erect, compact, agile, symmetrical, efficient, and enduring, he is properly the lord of the creation. Pre-existent nature was a prophecy of his coming. The physical creation rose step by step, platform upon platform, like a pyramid, whose apex is Man. Cicero says: "When you look upon a large and beautiful house, though you should not see the master and find it quite empty, no one can persuade you that it was built for the mice and weasels that abound in it." If we imagine some higher intelligence to have looked upon our globe at various periods of its formation prior to the appearance of man, he must have seen that this structure was as yet incomplete, that it could not be designed for the mere home of star-fish and lizards, that there must be some higher order of being for whom all this was preparing.

LECTURE III.

The Origin of Man.

THEORIES OF DEVELOPMENT.

THAT man is not coeval with the globe that he inhabits, but came into existence only in the last great age of geological time, that he crowned the series of organic life, that he was endowed with intellectual and moral faculties, and invested with dominion over Nature, are points upon which science attests the statements of the Biblical history. But a school of scientists deny that man was the immediate product of a new creation, and refer his origin to a *Law of Development* which they profess to trace in all organic Nature, working through secondary causes, without the intervention of a personal Creator. Various as are the theories of development, they all agree in ascribing the successive forms of life to secondary causes. As applied to Man, this doctrine demands a careful investigation.

The notion that man was somehow developed out of the Simia is not of recent origin. Eighteen centuries ago Pliny wrote, "Man is the being for whose sake all other things appear to have been produced by Nature;" yet he remarked, also, that "the various kinds of apes offer an almost perfect resemblance to man in their physical structure." Professor Huxley has made no advance upon Pliny in his statement that, "so far as structure is concerned, man differs to no greater extent from the animals which are immediately below him, than these do from other members of the same order;" but Huxley draws from this the inference, "that man has proceeded from a modification or an improvement of some lower animal, some simpler stock." This idea found a tangible expression in the early pagan mythologies. The god of flocks and shepherds among the Greeks was believed to be a compound creature, having the horns and feet of a goat and the face of a man. Their satyrs or forest divinities were creatures that blended the animal with the human. The fauns of Roman legend were supposed to mark the transition from the brute creation to man,—an idea that Hawthorne has finely wrought up in his "Marble Faun." Thus the question of man's development out of some lower type of creature does not lie between new discoveries of science and old dogmas of theology. The notion is as old as the oldest fables. Still, it deserves most candid consideration. We will first define precisely what the doctrine is.

The question is not that of a progressive order in the creation as a whole, but of the development of superior species from inferior by mere natural laws,—and especially the development of man from animals next below him in the scale of life. These two things must not be confounded. In the plan of the physical creation there are distinct traces of a progressive series in the types of existences. This is evident from organic remains in the strata of the earth. Professor Guyot has shown* that the formation of gases, of minerals, of water, —in a word, of the various constituents of inorganic Nature— must have subsided before life could begin. Also there was an *adaptation* of the physical condition of the globe to successive grades of life, which evinces purpose, plan, and therefore Intelligence. And there was an advance in systems. Thus we find growth in fishes, nutrition in reptiles, motion in birds, and symmetrical union in mammals; and

---

* Notes of his unpublished lectures on Man Primeval, delivered as the Morse Lectures in Union Theological Seminary, for 1869.

again the gradations of Matter, Life, Soul—the lower the *substratum* of the higher, but not its *source*.

But this advance in types is not necessarily the development of one out of another. On this point Mr. Darwin himself has been misunderstood and somewhat misrepresented. His speculations have no direct bearing on the Biblical doctrine of creation by a personal God. His mode of accounting for the origin of species does not dispense with divine causation. Mr. Darwin's theory is not that of spontaneous generation, for he maintains that "not only the various domestic races, but the most distinct genera and orders within the same great class, are all descendants of one common progenitor."

The development of the higher out of the lower assumes a gradation of orders, and the displacement of the lower in producing the higher; but Darwin teaches simply that the variation of species is induced by causes which already existed in the common progenitor. Neither does he teach origination by natural causes alone. Divergence by selection, resulting at last in prominent variations of type, he ascribes to natural causes; but the previous question, "How organic matter began to exist," he does not touch at all. He says, practically:—"Given the origin of organic matter, supposing its creation to have already taken place, my object is to show in consequence of what laws, or what demonstrable properties of organic matter, and of its environments, such states of organic Nature as those with which we are acquainted must have come about;"* in short, he is accounting for phenomena in species which have been brought to pass, as he alleges, by certain laws operating upon them since the original creation. On this point Professor Dana teaches that

---

* Statement of Darwin's views, by Prof. Huxley.

"species have not been made out of species by any process of growth or development, for the transition forms do not occur; that the evolution or plan of progress was by successive creations of species, in their full perfection. After every evolution, no imperfect or half-made forms occur; no back step in creation; but a step forward, through new forms, more elevated in general than those of earlier times; that the creation was not in a lineal series from the very lowest upward. The types are wholly independent, and are not connected lineally, either historically or zoologically. The earliest species of a class were often far from the very lowest, although among the inferior. In many cases the original or earliest group was but little inferior to those of later date, and the progress was toward a purer expression of the type. But Geology declares, unequivocally, that the new forms were new expressions, under the type-idea, by *created* material forms, and not by forms educed or developed from one another." *

To give one more authority on the same point, Professor Agassiz says: "Some have mistaken the action and re-action which exist everywhere between organized beings, and the physical influences under which they live, for a causal or genetic connection, and carried their mistakes so far as to assert that these manifold influences could really extend to the production of these beings; not considering how inadequate such a cause would be, and that even the action of physical agents upon organized beings presupposes the very existence of those beings. The simple fact that there has been a period in the history of our earth, now well known to geologists, when none of these organized beings as yet existed, and when, nevertheless, the material constitution of

---

* *Bibliotheca Sacra*, January and July, 1856.

our globe and the physical forces acting upon it were essentially the same as they are now, shows that these influences are insufficient to call into existence any living being.

"Nothing is more striking," he adds, "throughout the animal and vegetable kingdoms, than the unity of plan in the structure of the most diversified types. From pole to pole, in every longitude, mammalia, birds, reptiles, and fishes exhibit one and the same plan of structure, involving abstract conceptions of the highest order, far transcending the broadest generalizations of man,—for it is only after the most laborious investigations that man has arrived at an imperfect understanding of this plan; and yet this logical connection, these beautiful harmonies, this infinite diversity in unity, are represented by some as the result of forces exhibiting no trace of intelligence, no power of thinking, no faculty of combination, no knowledge of time and space. If there is anything which places man above all other beings in Nature, it is precisely the circumstance that he possesses those noble attributes without which, in their most exalted excellence and perfection, not one of these general traits of relationship so characteristic of the great type of the animal and vegetable kingdoms can be understood or even perceived. How, then, could these relations have been devised without similar powers? If all these relations are almost beyond the reach of the mental powers of man, and if man himself is part and parcel of the whole system, how could this system have been called into existence if there does not exist One Supreme Intelligence as the Author of all things."*

A strong *primâ facie* argument against the theory of development is found in the fact that, through all departments of the universe there are traces of invisible and immaterial

---

* Essay on Classification, Sections II. and IV.

Powers, that lie back of the phenomena that come under the direct cognizance of science, and are proximate causes of those phenomena. Thus, chemical affinity lies back of and produces the more important phenomena of inorganic matter; the principle of growth, which can not itself be analyzed or defined, produces vast changes in the vegetable kingdom; the principle of instinct influences many of the manifestations of animal life, and finally, a spiritual intelligence controls the actions of man. Does not this universality of invisible and immaterial powers point to a supreme spiritual Power back of *all* phenomena, and producing them?

It may be said, however, that the doctrine of development does not displace the personal Creator, but only removes to a greater distance the original act of creative power, which set in order the productive agencies whose results have been evolved in the successive types of existence. We come, then, to the immediate question, Was Man developed out of that which preceded him, and which was so manifestly a preparation for his coming? Suppose, for the sake of the argument, we admit that, as to his physical organization, man was but an improvement upon homologous structures in the animal kingdom, and that there was a progress through these successive forms up to the most perfect physical model, we shall then have provided only for the exterior case of the Man, by a plastic law of the Creator, and we must still refer the higher nature, the true spiritual humanity to God, from whom alone it could proceed. But we can not make even this concession concerning the lower physical organization of Man. President Hopkins,[*] in in an able discourse upon the principle of progress in the creation, calls attention to the distinction between a con-

---

[*] A Baccalaureate Sermon by Mark Hopkins, D.D., LL.D., President of Williams College.

dition and a cause. The universe, for instance, is built upon successive platforms of conditions, each platform being narrower than that directly beneath it, and the conditions being broader in their range of application than the thing conditioned thereupon. Thus gravitation, cohesion, and chemical affinity are all conditions of vegetative life; but these do not produce that life,—they are not its proper cause.

The Life is a new principle which enters from some other source, and lifts up these antecedent and necessary conditions upon a new platform for its uses. And so on through all the higher stages of existence, the end of the lower is the higher, but the lower ends without producing the higher, and has not in itself any power of producing the higher. Something not already in itself enters in to combine these conditions and produce the next higher plane. To the same effect is the teaching of Professor Dana:—that life and physical or inorganic force are directly opposite in their tendencies; that inorganic and organic Nature move in opposite directions; so that, on scientific grounds, we should conclude that physical force could not, by any metamorphosis, give rise to Life. Neither is there any authority from science to assert that Life itself is capable of more than simply living and reproducing itself. "Suppose the world to be in its condition of inorganic progress, we have no scientific ground for supposing that it could pass to a higher state, possessing living beings, by any parturient powers within. Or if Life exists, we still get no hint as to the evolution of the four sub-kingdoms of animal life from a universal germ; nor as to the origin of the Class-types, Order-family, or Genus-types, or those of Species, each of which is a distinct idea in the plan of creation. Nature, in fact, pronounces such a theory of evolution absolutely false. The perpetual pressence of Mind, infinite in power, wisdom, and love, and ever acting, is manifest in the whole history of the past."*

Professor Huxley, indeed, believes there is a "physical basis of Life," which underlies all the diversities of vital existence so that a unity of power or faculty, a unity of form, and the unity of substantial composition, pervades the whole living world. But granting that what he calls a "nucleated mass of protoplasm" is the structural unit of the human body, and that the body itself is a mere multiple of such units, still Professor Huxley admits the necessity of a *pre-existing living protoplasm* in order to the production of life; and for the origin of this he does not pretend to account. Indeed, he admits that we know nothing about the composition of any body whatever, as it is, and that chemical investigation can tell us little or nothing, directly, of the composition of living matter. Hence, of all the known forces and properties in the physical universe before man, we have no evidence that there was in them, singly or combined, a power that could have produced man as a living soul. The conditions of his existence were not the causes of his existence.

If man was produced by evolution from pre-existing organisms, where are the transitional forms? The change from the highest Simian type to the lowest human must have been gradual, and have extended over a long period. But no traces have been found of a creature intermediate between the ape and man, nor of a Simian tribe so far advanced as to fill up the gap. Professor Carl Vogt,† indeed, maintains that "microcephali and born idiots present as perfect a series from man to ape as may be wished for; and since it is possible that man, by arrest of development, may

---

* *Bibliotheca Sacra*, January, 1856.
† Lectures on Man; and Memoir on Microcephali, or Human-Ape Organisms.

approximate the ape, the formative law must be the same for both; and so we can not deny the possibility that just as man may, by arrest of development, sink down to the ape, so may the ape, by a progressive development, approximate to man." But this by no means follows. Exceptional cases of degradation from the superior to the inferior can not be held to prove a reverse law of progressive development from the inferior to superior. Vogt's reasoning is based entirely upon a few abnormal specimens of suppressed *human* development; whereas his argument requires that he should produce specimens of advanced *Simian* development, approximating humanity by slow but evident degrees. In the thousands of years since men and apes have lived side by side, the ape has made no advance toward the form, the habits, or the intelligence of man. Why has there been no lucky instance of a humanized ape, under the favoring conditions of human example, and with the supposed precedent of such a development given in the origin of man? And why has palæontology presented no specimen of the transitional ape, which had at least advanced to the level of idiotic humanity, resembling man in the organs of the body, though deficient in his manifestations of mind?

But while such resemblances as Carl Vogt has traced between abnormal specimens of humanity and the higher Simian types may give plausibility to a theory of development, there are, on the other hand, characteristics of man which so completely individualize him, and separate him from animals, as to neutralize the argument from resemblances.

Rochet has grouped these discriminating characteristics under five principal heads, a brief summary of which must answer my purpose for popularizing the subject.

(1.) *Man examined externally as regards form.* There

is not a single feature in the human face which, examined from an artistic standpoint, does not constitute a character of beauty and nobility foreign to the animal. Man alone has an expressive and intelligent physiognomy. This applies also to the body. The erect stature, the perfection of the hand and of the foot, are characters of the same value. The hand is especially characteristic. Man alone has a true hand; he alone uses this admirable instrument for creating the thousands of industrial and artistic masterpieces.

(2.) *The internal, sensitive, or moral man.* Man is endowed with a moral sensibility altogether unknown to the rest of organized beings. He loves or believes in things animals have no notion of. He possesses the feeling of the beautiful, the ugly, of wrong and right. He alone is conscious of the morality or immorality of his acts. Man alone has an idea of God, and is attached to him by feeling and intelligence.

Man alone of all animated beings forms a complete family. The animal takes life as it finds it, without any way modifying it. Man, on the contrary, takes life according to his will; for all the regions of the globe form part of his domain; and he can in a thousand ways vary the mode of his existence.

(3.) *Man considered as an active being.* Even in satisfying the lowest appetites, man differs from animals. He alone prepares his food by cooking it. Man alone provides himself with clothes to protect himself from the elements. When we treat of industry, instruments, and arms, the difference is enormous. Man possesses another important character,—intelligent speech.

(4.) *Man considered as an intelligent being—or the faculties of the human mind.* Animals possess a memory; but in them it is a faculty founded only on wants, personal utility, without any true notion of the objects; while in man, who, by means of language, conveys ideas, the facts of memory acquire great value. The animal possesses nothing analogous to the free-will of man. The animal entirely wants imagination, which for man is the charm of life, the consolation and the remedy for his evils.

(5.) *Man considered as a collective being.* The animal constantly loses territory which man gains. The day will arrive when there will be on the surface of the earth only such animals as are useful to man. Animality has no principle of cohesion in its members. Every animal lives only for itself. But men group together and combine their forces, and, although individually weak, they acquire an immense power. Man transmits his works and his conquests to his descendants. The animal perishes, and leaves only his skeleton behind.*

Now, these characters are *qualitative*, and serve to distinguish Man as a species. They belong to a plane so much higher than animal life that they must have been derived from a source above the laws and conditions of that life; they answer to and verify the place assigned to man by the Mosaic account of his creation; that he was made in the image of God, and invested with dominion "over the fish of the sea, and over the fowl of the air, and over the cattle, and over all the earth, and over every creeping thing that creepeth upon the earth." The adaptation of man for this supremacy over Nature is marked by that feature of his physical structure which Professor Dana has happily termed *cephalization.* "The head of an animal being the seat of power, containing the principal nervous mass, and the various organs of the senses, it is natural that among species rank

---

* These views of M. Rochet are condensed from the *Bulletins* of the Paris Anthropological Society, and published in the London *Anthropological Review*, April, 1869.

should be marked by means of variations in the structure of the head; and not only by variations in structure, but also in the extent to which the rest of the body directly contributes, by its members, to the uses or purposes of the head." Now, in man, the organs of digestion, of locomotion, and the like, are reduced to the *minimum* of the demands of a rational creature, while "his nervous system stands vertical, with the brain at the summit, and that brain nearly treble the size of the brain of a gorilla." The body in all its parts is placed directly under the domination of the head, and is fitted for head-uses. "The superiority of man to other animals has long been recognized in the structure of his *hand*, which is so wonderfully fashioned for the service of his exalted nature; in his *erectness of form*, which seems like a promise of a world above, denied the animal, which goes bowed toward the earth; in his *face*, which is made not only to exhibit the inferior emotion of pleasure, through the smile or laugh, but—when not debased by sin—to move in quick response to all higher emotions and sentiments, and calls for sympathy, as though it were the outer film of the soul itself; in his *speech*, which is the soul in fuller action wielding its power in force on other souls. We now perceive that these characteristics are outer manifestations of a structure whose elevation for the uses of the brain is in accord with man's greatness of intellect and soul. Thus living Nature, as with universal acclaim, bows before man its visible head. Man, the offspring, not of Nature, but of God, can not be brought within the plane of a material development without destroying all that is distinctive in Humanity." His dominion over Nature will be set forth more at length in a subsequent lecture. But we can not close this train of thought without a grateful recognition of the nobility and grandeur of our Humanity as first conceived in the design of God! There are some who object to the Biblical view of man, that it is degrading, that it makes no recognition of that dignity of which he is conscious, that it puts upon him no such honor as science accords to him in the creation. So far from this, it is the Bible that puts honor upon man in the record of his creation.

It is not the Bible that traces the origin of man back to the monkey or the trilobite;—this makes him the child of God, created in his image, for his companionship and his glory. True, the Bible represents man as fallen and degraded in character, but this by his own act, because God had made him a being of voluntary powers, which powers he perverted to his own degradation;—but, nevertheless, by reason of these very powers, he is capable of recovery and restoration to his original place and destiny as the offspring of God. As the highest organization upon the globe he inhabits, he is the crowning excellence of the creation. But this organic perfection is a small part of the Creator's ideal in man. When, after all his other work, God said, "Let us make man in our own image, after our likeness," He set him apart from all other creatures in a sublime pre-eminence, and put the seal of divinity upon him as an intelligent soul; and then, as if to represent Himself upon the earth, He crowned man with glory and honor, and set him over all the works of His hand. No theory of development, no speculation of philosophy, no dream of poet can place man upon such a pinnacle of honor as that where God set him at the first. He has thrown himself down from that position of dignity by self-will, self-worship, the love of the creature,—by knowing, willful, daring disobedience of God. Man is not a poor, struggling creature, just breaking away from the fellowship of brute beasts and making fitful endeavors after a higher life;—he is a fallen creature. The image of God, a little lower than the Elohim, he has debased himself to the level of creatures of the earth and earthy. Take away sin from man, and he would no longer grope after an affinity with brutes, but feel again his fellowship with God. As Tholuck finely says, "We are feudal servants, holding our title over the lower creation by grant from the Creator and Lord of all. But, elated by arrogance, the feudal servant has rebelled against his feudal lord. We ought to consider ourselves servants, but set up ourselves as independent lords of creation. We ought to be the priests of God, re-offering to Him, and using for His glory, whatever His creation has provided for us; but have become idolaters, worshiping the idols of our own selves. It is one of the effects of that rebellion, that our royal scepter became broken, and that only a fragment of it remained in our hand; for our present knowledge and power are but poor fragments of the glory which we were originally destined to enjoy."* That glory man can not regain by material means. No progress in the physical sciences can ever restore him to his forfeited position. The soul is the true seat of dominion, and his restoration must come through the renovation of the soul.

Suppose that, with infinite pains and daring risks, one climbs to the summit of Mount Blanc; he can not stay there, — he must either perish with cold or die from the difficulty of breathing at that height. He finds himself encompassed also with clouds that intercept his vision, so that only by rare glimpses can he see farther than from many a lower peak. He can have only for a moment the vain satisfaction of having outclimbed his fellows, and must descend again from this chilling height, with no new dominion over Nature, to share the common lot of men. But

---

* Tholuck, "Commentary on Psalm viii."

with a soul renewed to holiness, he can rise to Alpine heights of vision and of glory, higher and yet higher, commanding at each ascent some wider prospect of truth, inhaling a purer atmosphere, gathering strength as he rises for yet loftier attainments, evermore rising toward God, his source, his center, and his all. Of Him, and through Him, and to Him are all things: to Him be glory forever. Amen.

LECTURE IV.

Man's Dominion Over Nature.

Does Man belong to Nature as begotten of it, included in it, concluded by it? or does Nature belong to Man as his original birthright, his temporary habitat, his ultimate dominion? Is it true, as some physicists affirm, that Man is just the latest outcome of Nature's efforts at improving upon her own experiments in organic life—the treasured selection of some accidental variety of birth in a Chimpanzee family? or, as say others, that he is "but the last term of an innumerable series of organisms which has been slowly evolving under the domination of the same law?" or that Man, whenever and however he began to be, is "under the absolute control of physical agencies," cradled by Nature and molded at her will? Have we done with Personality, done with Consciousness, done with Liberty—except as a name to fight for—done with Progress, save in the fixed and narrow groove of physics and statistics—which, after all, is not progress, but the rotation of natural forces in an ever-returning cycle?—have we done with Spiritual Powers, and with Causes both intelligent and final?—have we done with the Deity save as impersonal fate or law, and having done with God have done also with Man,—for whom there is neither dignity, worth, nor hope if there be no God? Quite otherwise would I seek the solution of the problem of life. I find in it three factors, co-operative but not co-ordinate:—God, Man, and Nature. What, then, is the normal relation of Man to Nature? or, if you please, what are the mutual relations of Man and Nature, the two mundane factors in the problem of life?

Without question we yield to Nature precedence in the order of time. Nature was before Man. Through immeasurable æons the *processus* of her phenomena, in all their varied beauty, sublimity, and terror, had moved on with no human spectator to observe them. The upheaval of the continents; the slow subsiding of the seas; glaciers and icebergs, volcanic fires and steamy mists—hot, cold, moist, dry, striving for mastery "o'er many a frozen, many a fiery Alp;" gigantic flora blooming and decaying; monsters of reptile and animal life, the spawn of chaos and night; all these had been, and had left their record upon the surface of the globe; —inorganic nature, organic nature, life vegetable, insect, animal, *all* had passed on and on through timeless epochs of duration without one trace of Man.

And when we reflect with the geologist, that "from the inconceivably remote period of the deposition of the Cambrian rocks the earth has been vivified by the sun's light and heat, has been fertilized by refreshing showers and washed by tidal waves; that the ocean not only moved in orderly oscillations regulated, as now, by sun and moon, but was rippled and agitated by winds and storms; that the atmosphere was influenced by clouds and vapors, rising, condensing, and falling in ceaseless circulation," and yet that while Nature was thus established in her ordinances, through the long, long ages from the Cambrian to the Post-Tertiary there was no human organism, we are impressed not only with the recency of Man's origin, as compared with the whole duration of the globe, but with his physical insignificance upon the scale of the universe. In this view we concede the grandeur of Nature, in her antiquity, her forces, and her laws.

Geology, as we have seen already, teaches that Man began to be. But it also teaches that prior to Man's appearance "the material constitution of our globe, and the physical forces acting upon it, were essentially the same as they are now;" that phosphate of lime, iron, and albumen had then the same properties as now; that heat and electricity had the same vitalizing power, and that these materials and forces were then as now the same in their combinations and effects. But we can take these materials and forces into the laboratory, and there measure, analyze, and combine them, and ascertain just what they are capable of effecting, and that they are *not* capable of originating the human organism or of producing human life, even when directed by the science and the ingenuity of Man with an analysis of the human subject before him. Something more than mere physical Nature, even after long ages of her evolution, is required to account for the appearance of Man upon her stage.

And we may go farther. Not only are mere physical forces inadequate to originate life, but there is much to warrant the position of Professor Owen, that these forces are in antagonism with life, and tend to its destruction; that every living organism has "to maintain a contest against the surrounding agencies that are ever tending to dissolve the vital bond, and to subjugate the living matter to the ordinary chemical and physical forces." *

A change of climate, a wet or a dry season, a wind, a flood, is not only largely destructive of life, but such mutations long continued may extirpate whole species; so that life depends upon the self-adjusting power of the individual in respect to the hostile and destructive forces of Nature. The eminent authority just cited reminds us that

---

* "Palæontology;" see Note at close of this Lecture.

"with life, from the beginning, there has been death. The earliest testimony of the living thing, whether coral, crust, or shell, in the oldest fossiliferous rock, is at the same time proof that it died. Hence the operation of *creative* force has been limited to no one geological epoch; but palæontological research has established the axiom of the continuous operation of the ordained becoming of the species of living things." Not natural evolution, but creative interposition upon the plane of Nature, is the lesson of the record of the rocks.

In the present stage of science, I may safely lay down the postulate, that Man had a beginning, and that Nature is not proved adequate to have caused that beginning. He appeared upon the plane of Nature with an organism that Nature fails to account for, and with powers for which Nature furnishes no precedent.

In the preceding lecture I have granted all that can be fairly claimed in the facts of science by the advocates of the notion of development under whatever phase,—especially the two cardinal facts of a progressive order in the types of existence from the zoophyte up to the mammal, and of the homologies among different classes of vertebrates, from fishes up to Man; yet the theory of evolution remains a *theory*, or rather an hypothesis, transmitted from the oldest pagan mythology, and is no more an established dogma of science in the pages of Darwin and of Huxley than in the pages of the "Marble Faun." Nay, is not Hawthorne even nearer the truth when he ascribes the transformation of the mute mystery of the animal in the Faun into the consciously human in Donatello, to a crime prompted by the passion of love, that awakens at one stroke intelligence, conscience, guilt, death? But artistic and scientific Fauns and fancies aside, the vital question is whether Man was *created*, or whether, like Topsy, he *growed;* whether he has simply a

SERIAL PROGRESSION NOT EVOLUTION.

"place in Nature" as one of her series, or a position *over* Nature by reason of personal prerogatives and powers. We must be careful not to confound things so widely distinct as progress in the series of animal life and the evolution of higher species from pre-existing organisms.

It is a most unscientific defect of the theory of development, that it ascribes to known causes unknown effects. The causes are before us; we can measure exactly their power, can trace minutely their operations, can observe their effects. Yet effects which they have never been known to produce, and which sustain no natural nor logical relation to these causes, are now assumed to have proceeded from them by some mysterious law of evolution in the past, which has never renewed its activity for the gratification of a human observer. The hypothesis rests upon assumption. It may be illustrated by an analogy.

In a great pottery one sees common earthen vessels of coarse grain and uncouth shape stored in the basement; and above these a pure white glazed ware of plain patterns; and above this vessels of artistic forms and ornaments, but made of the same simple materials; and above these, again, vases of porcelain or of *terra cotta* of the most delicate structure in their material, and decorated with the most exquisite touches of art: the Wedgwood vase, that rivals the choicest specimens of Etruscan antiques; the Sèvres china, that vies with the most curious workmanship of Japan;—but, though there is in these various structures a progress in workmanship and design, one class is in no wise an outgrowth of another; but the progressive series witnesses for the inventive skill of the artificer. A progressive order in structure does not prove the development of each more advanced individual or class in the series out of that which next preceded it. The fact that there was an order and a progress in the forms of organic life as these were brought upon the stage of existence, affords no proof that there was a development of one form out of another by some natural law; this may only unfold the intelligent plan of the Creator in his works, the order of the Creator's acts—the power of the potter over the clay. Mere homology of structure does not prove evolution. In the factories of Lowell one sees carpets of divers quality, figure, and texture woven by the same motive-power; but an intelligent will devises the pattern and adjusts the loom to that combination of materials which makes the difference between them; and no principle of development, no accidental or natural variation will account for that difference. And so in the world of life,—Nature, acting like a vast power-loom, may work up her materials upon some general plan of structure with varieties of form; but the loom does not originate either the structure or its varieties; it simply works up the materials that are put into it, according to the patterns devised and set by the creative mind. Plato was right in counting the divine ideas the real substances, and those conceptions which originate in the intelligent will of God, Nature, acting as His power-loom, *must* work up according to the pattern. She can not, of herself, pass from one to another, for Nature is under law to the will of her Creator. Hence, as a leading naturalist has said, "the resemblances between the skeletons of Man and the Apes may, to the uninitiated in science, appear to make the transition by development feasible, yet they are of no weight as argument, since the question is as to the *fact* whether, under Nature's laws, such a transition has taken place as the gradual change of an ape into a Man, or, whether apes were made to be, and remain, apes?" There is no evidence whatever from any half-and-half specimens, or from any traces in organic remains, of such a gradation from the gorilla up to the human organism. The *gap* between the two is still immense; and homologies are no part of the fact of a transition from one to the other in the remote ages of the past. The theory of such a transition will not explain the amazing differences between Man and the lower creation. The strongest advocates of gradation do not pretend that any remains have been found of a human being intermediate between men and apes; or that take our race down appreciably nearer to that lower form of animal existence. A union now of the so-called transmuted species with its original could only issue in a *monstrum horrendum, informe, ingens.*

The Duke of Argyll has stated with clearness and force the objection to the theory of development from the absence of intermediate links in those portions of the geological record which are the most consecutive and complete. "The Silurian rocks, as regards oceanic life, are perfect and abundant in the forms they have preserved, yet there are no fish. The Devonian age followed, tranquilly, and without a break; and in the Devonian sea, suddenly, Fish appear,—appear in shoals, and in forms of the highest and most perfect type. There is no trace of links or transitional forms between the great class of Mollusca and the great class of Fishes. There is no reason whatever to suppose that such forms, if they had existed, can have been destroyed in deposits which have preserved in wonderful perfection the minutest organisms."[*] The same writer argues that "the human frame diverges from the structure of the brutes in the direction of greater physical helplessness and weakness—a divergence which it is most impossible to ascribe to mere Natural Selection. The unclothed and unprotected condition of the human body, its comparative slowness of foot, the absence of teeth adapted for prehension or defense, the same want of power for similar

---

[*] "Primeval Man," p. 45.

purposes in the hands and fingers, the bluntness of the sense of smell, such as to render it useless for the detection of prey which is concealed—all these are features which stand in strict and harmonious relation to the mental powers of Man. But, apart from these, they would place him at an immense disadvantage in the struggle for existence. . . . The lowest degree of intelligence which is now possessed by the lowest savage is not more than enough to compensate him for the weakness of his frame. If that frame was once more bestial, it may have been better adapted for a bestial existence; but it is impossible to conceive how it could ever have emerged from that existence by virtue of Natural Selection. Man must have had human proportions of mind before he could afford to lose bestial proportions of body. If the change in mental power came simultaneously with the change in physical organization, then it was all that we can ever know or understand of a new creation.*"

But it is claimed that if Man be not a product of Nature by a progressive law of evolution and selection in living organisms, he is yet so completely under the control of physical circumstances that Nature determines his character by her conditions, and rules him by her laws. That physical geography affects the characteristics of race is patent all over the globe; and a sound sociology must make account of soil, climate, vegetation, mines, mountains, rivers, seas, as well as of intellectual and moral phenomena, in estimating the qualities and the prospects of a people. But the question remains whether the influence of physical conditions upon human life is so uniform and absolute as to amount to a determining cause? or does Man possess an essential quality of *dominion*, which makes him the proprietary of physical Nature, how-

---

* "Primeval Man," pp. 66–70.

ever he himself may be impressed or modified by its conditions? so that—as Marsh expresses it—"though living in physical Nature, he is not of her, but belongs to a higher order of existences than those born of her womb and submissive to her dictates."*

We will here concede, for the sake of argument—though the data are not sufficient to authorize the conclusion †—that Man began his existence on the earth at the low stage indicated by the relics found in the valley of the Somme and the caverns of Liege, and by the pile-habitations in the Swiss lakes. But this crude stone-period shows us Man — the most dependent of the animal creation — nevertheless subduing Nature to his uses; an artificer inventing tools, and planning houses, first converting stone into arrow-heads, lance-heads, axes, hammers; then inventing bronze, and applying iron to his purposes by the laws of heat; and by what Humboldt aptly styles "the flexibility of his own nature," making physical Nature under every form and in every clime to minister to his wants and pleasures. In the stone-age, he was the builder and the inventor; the bridge, the aqueduct, the railway, and the telegraph were in him *in posse*, for the dominion of the world was his, and he had but to "fight it out upon that line." This feeble, timid creature came to share at first the caves and forests of the beasts—so say, at least, this school of archæologists;—but Man advances, and the beasts disappear or are tamed. Whole races of animals, once contemporary with Man, have become nearly or quite extinct, while he, so inferior to them in size and strength, has multiplied till he has overspread the earth. By his progress the wilderness has been subdued and made a fruitful field;

---

* G. P. Marsh, "Man and Nature."   † See Lecture V.

marshes have been drained; deserts reclaimed by irrigation; and climates at first hostile and deadly, have been mitigated or counteracted in their effects by human care and skill. The Nile yields up at last the mystery of its sources; the Arctic can not long hide its secrets; the wild Atlantic consents to be bound by cables to either shore, and is linked to the Pacific by iron bands that span a continent. Man subordinates the whole creation to his own uses. He gathers the fruits of the earth, the products of its mines, the treasures of the sea; he employs the subtile agencies of light, the powers of heat and of motion, the fearful velocity and energy of the lightning. According to his latitude and his wants, he employs the reindeer, the dog, the horse, the ox, the buffalo, the camel, the elephant for transportation, or extracts from fire and water the motive power of steam. He gets light from the fat of sheep and oxen, the blubber of the whale, the coal of the mountains, the resin of trees, the rivers of oil in the bowels of the earth. He clothes himself with fabrics woven from the skins of animals, the plumage of birds, the pods and fibers of plants and trees. In a word, he makes all Nature contribute to his use, his comfort, his taste, his pleasure, and this by the mere brain-power lodged in him as lord over the creation. If one would realize man's position over Nature, I know not where to study it to more advantage than in the Smithsonian Institute and in the Patent Office at Washington:—in the one you have an exponent of Man's comprehension of Nature through science; in the other, of his combinations and adaptations of Nature through invention. The one shows his mastery of the principles of Nature, the other his mastery of the forces of Nature. At the Smithsonian, you see in its museum how Man has studied, subdued, and classified the whole animal kingdom; in its laboratories you see how Man has investigated the secrets of Nature, and learned to measure and apply her forces; in its library, you see what knowledge Man has amassed concerning the world in which he lives; and in every department you see his brain stamped upon all as proof of his lordship over all.

In the Patent Office, you see all material substances and mechanical powers combined and adjusted in countless forms of ingenuity for the benefit of Man. How many contrivances for heat, for light, for motion, for clothing, for building, for navigation, for printing, for safety, for defense; how many devices of art and taste for ornament and pleasure; how many varieties in the application of the same principles and materials—yet every one of the myriad of inventions in these long corridors crowded with models is the creation of the human mind. No gorilla ever took out a patent, or made any improvement upon the condition in which he was born; no man-like ape ever developed out of Nature anything beyond what his instinct taught him at the first. No law or process of Nature ever produced a machine. All these materials, all these agents and forces, all these possibilities of adaptation and usefulness, lay slumbering in waste, in quarries and mines of the earth, in the bosom of unconscious Nature, until Man, standing upon Nature and above it, wrought them into shape for his own use and made them his servants. Who can walk the corridors of the Patent Office and deny that Man is lord over Nature? Where among these models would you arrange Man as having a place *in* that Nature over which he reigns supreme?

Man is the *user* of Nature. Always and everywhere is Nature used by him for his own ends. But, as Socrates argues with Alcibiades, "he who uses, and that which is used, are different"—the currier who uses the cutting-knife different from the instrument he uses, and different also from the hands and the eyes which he uses in his work. Nor can we rise above the grand simplicity of Socrates' definition of a man. "The Man is that which uses the body:—now, does anything use the body but the mind? Is not the mind, therefore, the Man?" We answer with Alcibiades, "The Mind alone."

To clinch the argument, here comes in the fact that Man's conquest of Nature is in the ratio of his spiritual development. The higher civilization upon our globe is nearly coterminous with Christianity. This correspondence is not accidental, but illustrates the law of Man's nature declared at the beginning. As the offspring of God he was invested with dominion over the world; by the fall he lost his spiritual supremacy, which is again restored through the new life in Christ.

I do not forget that the Phœnicians and the Egyptians had arts now lost to us, whose products are curiosities in the scientific museums of the nineteenth century. I do not forget that the arch once called Roman is as old as Egypt, and the aqueduct as old as Tyre; that glass was molten on the shore of Phœnicia, and linen spun upon the banks of the Nile; that modern architecture is so largely a copy of the schools of Greece, and that in the modeling of marble and bronze, antiquity is still our teacher. In the time of Solomon mankind had already "found out the knowledge of witty inventions," and "of the making of books there was no end." Yet the records and monuments of antiquity show us, in the main, the multiplication of works of physical strength rather than inventions of practical utility. A reaping machine traversing the delta of the Nile, a steam pump to irrigate the soil after the yearly inundation, would have done more to assert Man's dominion over Nature than the piling up of stones into pyramids and the carving of tombs and temples from the solid rock. Yet the mechanic arts of ancient Egypt, the arts of practical life as depicted upon her monuments, were in a comparatively unscientific stage long after the pyramid of Cheops was pointed to the sky. Whatever the land of the Pharaohs now possesses in scientific improvements applied to the uses of practical life, is not an inheritance from her ancient civilization, but a benefit imported from Christian lands.

In countries where the influence of Christianity is marked, we notice a rapid and constant progress in Man's dominion over Nature. In all such countries improved methods of agriculture are making the earth more and more tributary to the sustenance of Man. Science has instructed the husbandman in the qualities and capabilities of different soils, and the means of enriching them so as to enhance their productive value. Thus the increased productiveness of the earth is made to keep pace with increased density of population. Improved implements of husbandry economize the labor of Man, and develop more fully and perfectly the fruits of the earth. A subsoil plow, and a common force-pump for irrigation in the dry season, would cause the plain of Sharon in Palestine once more to blossom as the rose. But the Arab fatalist continues to scratch the earth with a stick, and water it with his crazy creaking bucket; to tread out grain by the feet of cattle, and winnow it by a shovel in the wind. There is no progress in the subjugation of Nature visible under the rule of Mohammedanism; hardly more is to be seen where Christianity has degenerated into a superstition and a name. But wherever Christianity exists in tolerable purity and power, there the earth seems gladly to acknowledge the dominion of Man. Christianity inspires a philanthropy that makes the elevation of the working classes its study and care, and that infuses into political economy the principles of the highest morality.

Under the influence of Christianity, Man regains his dominion over Nature by the conquest of physical evils. Such evils are better comprehended and therefore more easily subdued. The lightning and the storm, once the terror of the pagan mind, are measured in their nature and effects, and met with a fearless will and with adequate defenses. The marshes that once bred disease and death are drained and converted into meadows and orchards. The jungles that once harbored wild beasts and venomous reptiles are cleared and converted into fruitful fields. The desert is made to pour forth water from Artesian wells. No evil is longer deemed insurmountable to Christian civilization. Livingstone goes forth anew with the implements and appointments of that civilization to subdue and renovate the barbarism he has explored in the wilds of Africa. The steamboat shall penetrate the waters of that vast continent, and the clearings of civilized men shall sweep pestilence from its shores. The railroad which has already conquered the Egyptian desert, the Rocky Mountains, and the salt wastes of Utah, shall yet conquer and reclaim the vast wastes from the Mediterranean to the Euphrates. Whereever a Christian civilization advances, its paths drop fatness; the valleys are covered with corn, the hills are girded with joy.

Under the influence of Christianity, Man regains dominion over Nature through the development of the interior resources of the earth. There were quarries and mines in ancient times. The Egyptians worked veins of copper in the peninsula of Sinai before the days of Joseph. The gold of Ophir and the pearls of the Indian Sea enriched the coffers of Solomon. But chance or the diviner's rod discovered such treasures, and unskilled labor could only scratch their surface. The science of Geology has been born within our day; and now there is not a Christian state which does not make a geological survey a guide to the resources of mineral wealth hidden beneath its soil. The intelligent and profitable working of coal and iron, and of the various minerals and ores which the all-provident Creator has stored against Man's time of need, is a result and an attendant of Christian civilization. The semi-civilized nations of the Eastern world universally look upon the Franks—the representatives of European Christianity—as having a special insight into Nature.

Under the influence of Christianity, Man regains dominion over Nature through the discovery of occult principles and their application to practical uses. In this field Man has achieved his greatest triumphs over Nature by coming at the very secrets of her power, and turning these to his own account. How long had Nature hidden these secrets from the eye of Man! From the day when first the waters gathered into vapor above the seas, there has existed a force which could convert the sea and the land into a highway of commerce and of travel for Man. From the day when he first brought fire into contact with water, he had in these opposing agents a power which he might employ to subdue Nature in every form. From the day when the first stroke of lightning on a tree terrified Man with its instantaneous power, he had within his reach an agent that could make his own thought instantaneous and almost omnipresent through the world. Yet Man who was invested with dominion over these forces has lived, and labored, and suffered, and died in all the generations of six thousand years, without once suspecting that the keys of his sovereignty over Nature were in the vapor steaming from his kettle, and in the magnet which he knew only as a toy. Vainly did he knock at the door of Nature and summon her to disclose her secrets; she was dumb to force and to command. But when once he learned the "Open Sesame," her doors flew back on golden hinges, and not Aladdin's lamp disclosed such jewels as in the crown of glory and honor which the Creator had here laid up for Man. "Thou *madest* him to have dominion over the works of thy hands; thou hast put all things under his feet; all sheep and oxen, yea and the beasts of the field, the fowl of the air, and the fish of the sea, and whatsoever passeth through the paths of the seas."

This discovery of the occult principles of Nature, and the application of those principles to the service of Man, is the highest triumph of the spiritual over the material. A principle or law of Nature is the will of the Almighty impressed upon and acting through material forms. Science teaches us that these occult principles of Nature are not spontaneous, self-originated, independent, self-sustaining forces inherent in matter as such, but when traced back into their remotest confines are still *laws* impressed upon matter by a planning mind. It teaches, moreover, that these laws pertain to and are to be sought in the *molecules*, the minutest particles of which bodies are supposed to be composed. And when we begin to touch upon such laws, we come into an awful nearness to Him who framed them. It is as if one groping in a cavern found himself upon the verge of a precipice, and reaching forth at that instant should lay hold on something in the dark, and feel the pressure of an invisible hand. Here, in the arcana of Nature, God takes us by the hand, and giving us the laws by which He governs the world, He gives us dominion over the works of His hands. How marvelous has been the growth of that dominion from the day when a poor Frenchman was shut up in the madhouse for saying that he could move ships and carriages by steam, to the day when Stephenson ran a locomotive sixty miles an hour! How great the march of that dominion from 1744, when Franklin drew down the lightning by his kite and key, to 1861, when New York and San Francisco were brought together by an instantaneous flash of thought!

Some would ascribe this superiority in science and invention, which is so marked in modern Christian nations, to what is termed superiority of race. But were ever races, as such, superior to the Roman and Greek races, that gave law and art, eloquence, poetry, and philosophy, to the world and to time? Besides, where and what were the present superior races of men before they were touched and invigorated by Christianity? What were Gaul and Britain and Germany but the seats of a barbarism that the conquering Romans scorned? This wondrous development of Man in the way of material progress was preceded by the moral renovation of these dominant races through the Gospel; and it belongs emphatically to the era of a free, spiritual, and practical Christianity, ushered in by the Reformation. The renovation of Man's spiritual nature by Christianity has restored him to his original supremacy over the physical world, as the offspring of God.

The contemplation of Man as a builder and inventor, and therefore the superior of Nature, even in his material wants and conditions, brings him before us as an Intelligence, and therefore as a Spiritual Power. When we rise from the sphere of instinct into that of intelligence, we leave the range of material laws for that of spiritual powers; and the position of the brain in Man for guidance and control, with its size and power, and its improvable quality, marks an impassable distinction between Man and the whole animal creation. "In the living beings of former ages," writes Professor Dana, "there had been intelligence and a low grade of reason—affections as between the dam and her cub, and the joyousness of life and activity in the sporting tribes of the land. But there had been no living soul that could look beyond time into eternity, from the finite toward the infinite, from the world around to the world within and God above. This was the new creation—as new as when life began; a spiritual element as diverse from the life of the brute as life itself is diverse from inorganic existence." And this new life was typified, as we have seen, in the cephalized structure of Man,* who from the first appeared as a creature formed in all his parts for the service of his brain, and thus for rational dominion over inorganic matter and mere physical life; a creature who,

> "endued
> With sanctity of reason, might erect
> His stature, and upright, with front serene
> Govern the rest, self-knowing; and from thence
> Magnanimous to correspond with heaven."

Man has been defined as an Intelligence served by organs; and his reasoning intelligence is a characteristic that separates him from the brute creation by a chasm that they can never cross. The contrast is most striking when the human mind is directed to a point where the instinct of an animal is exhibited in the highest perfection. Only by the refined and severe method of the calculus was it ascertained that to secure the most room and strength upon a given space, with the least waste of material, the builder must adopt the exact angles which the bee forms by instinct. But how much

---

* Cuvier's great discovery, which Prof. Owen styles the "law of the subordination of the different organic characters to the condition of the whole animal," finds its highest example in the subordination of Man's body to his brain.

greater the mind of Newton that grasped the principles, and defined the laws, and gave the rules of calculation, than the instinct of the bee in doing its work! How much greater the mind of Michel Angelo shaping St. Peter's to his thought, and then crystallizing the conception into stone, than the instinct of the bee building its cell! Whence came the mind of Newton, the mind of Michel Angelo? Was this developed upward from the instinct of the bee? or was it a created intelligence, the offspring of God? And what shall we say of this MIND of Man?—its power of reasoning, which grasps the facts of the external world and the truths of the inner world of consciousness, and weaves them into consecutive chains of ideas, and builds up fabrics of thought that will stand though the physical universe shall fall?—the Mind which hides itself within its net-work of nerve and sinew and muscle, like an invisible spider, alive to the least touch or approach from without, quick to seize upon and appropriate as its food whatever comes within its range, throwing out new filaments to bind each floating atom of the real world, and then spinning from its mysterious depths a new world of thought and imagination, of ethereal texture and prismatic beauty, itself the living center of the whole? What shall we say of this Mind that, from a few arbitrary characters and a few articulate sounds, constructs a language that expresses thought, that stirs emotion, that kindles passions or allays them—language that makes the printed page glow with the fire and beauty of poetry, that makes the air pulsate with the throbs of eloquence?—this Mind that from a few arbitrary figures, that you may count upon your fingers, constructs the abstract science of mathematics, by which it weighs the mountains in scales and the hills in a balance; by which it measures the velocity of light and the distances and magnitudes of the stars?—this Mind of Man that, with unfaltering confidence, determines by mathematical law that the equilibrium of our solar system demands the existence of another planet yet unseen, then points the telescope and finds it where it ought to be?—this Mind that takes the wings of the morning and out-travels light; that flies backward to the beginning and forward to the unknown; that counts all time and space its home, and dares look forth upon the Infinite? From a few letters of the alphabet Homer made a poem whose rhythm still beats upon the shores of Time, while the sea washes a desolate beach where Troy once stood; Plato gave shape to thoughts that live, while Athens is falling to decay; the creations of mind survive, though temples and pyramids perish; and though the heavens should pass away, and the stars be seen no more, the system of mathematical order and beauty that Newton formed from a few abstract lines and numbers, would remain for the admiring contemplation of the Mind, overarching it with a firmament of its own. This Mind of Man, with its powers of Reason, Imagination, Memory, Will,—with its hopes and fears, its joys and loves,—this Mind that *knows itself*, and that dominates all matter and all life without itself,—can it be less than the immediate offspring of God? If Man be over Nature as a power, is not God more than Nature, more than law, more than fate? Is not Man himself a proof of the *super*natural?

Man, whose physical origin can not be traced to any evolution of natural law, whose rudest beginnings of life were an assertion of his dominion over Nature, whose functions as an Intelligence ally him to the realm of spiritual powers, is separated yet more decisively from the control of Nature in the sphere of Consciousness. That is an untenable and absurd monopoly of the term science that would restrict it to physical phenomena, and would treat of these as the only realities in the universe. Indeed, we can have no certainty concerning that which is without, save upon the assumed certainty of that which is within. All knowledge from the widest circle comes back at last to the knowing subject as its pivot; and the physicist who reposes all his faith upon Nature, would have the unscientific layman receive Nature itself upon faith in his observation and veracity—in other words, faith in *himself* as an intelligence. Now, this selfhood is the essence of the Man, the conscious person, the Ego, who knows himself, and knows that he knows; and within himself there is a domain of science, of nobler phenomena, and of no less certain determinations, than the physical universe. "When I was young, Cebes," says Socrates, in the Phædo, "it is surprising how earnestly I desired that species of science which they call physical. For it appeared to me pre-eminently excellent in bringing us to know the causes of each, through what each is produced and destroyed, and exists. But happening to hear some one read in a book, which he said was of Anaxagoras, that it is Intelligence which is the parent of order, and cause of all things, I was pleased with this cause, and it seemed to me to be well that Intelligence was the cause of all, and I considered that, were it so, the ordering Intelligence ordered all things, and placed each thing there where it was best." The science that deals with Intelligence as its subject is of a higher order than that which deals with physical phenomena and their laws. The science of Mind is higher than the science of matter, and the science of Morals is higher than either, in the nature of its subject and the grandeur of its results.

Nor can we concede to physicists that theirs is peculiarly the science of certainties. Personality and free-will, given in consciousness, are as certain as are the mountains. Right and wrong, truth, justice, moral law, are as certain as are any facts in the physical world. When, therefore, Dr. Draper or

Herbert Spencer would persuade me that I am shaped and governed as an atom by purely physical causes, I assert my conscious personality and free-will. If he seeks to contravene these by physical laws, I still *assert* them, and defy him to set them aside. For if the Ego does not exist as a conscious subject, the perception of the non-Ego, which is Nature, is an impossibility; or, as Hamilton expresses it, "once Consciousness is ruined as an instrument, Philosophy is extinct."

The materialist insists that I shall believe only that which can be tested by the evidence of the senses, and reduces the operations of mind itself to manifestations of the physical organization with which it is connected. But in recording observations and making experiments that extend over a considerable period of time, he assumes his own identity through memory and consciousness—his personal existence as an intelligent observer—and he demands of me that I shall accept the results of his observations upon faith in his intelligence, his competence, his accuracy, his fidelity;—in a word, he demands of me faith in human testimony concerning that which in theory he holds should be accepted only upon the strictest scientific evidence. Now the certainty of the facts of consciousness which the materialist tacitly assumes, while he rejects them in theory, may be no less conclusive and absolute than the certainty of physical facts observed by the senses.

I *know* that I am, that I think, that I will, that I am free. I know that there is a Right, a Justice, a moral Law. I accept whatever facts the materialist brings me from his varied and profound researches in the domain of physics; but when he seeks to bind me with these as with chains, I say to him, There are other facts also, as certain as yours, and nobler, grander far than yours; these I know, and these you too would perceive, if, like Socrates, you would cease to be enamored of mere phenomena, and rise to the sphere of Intelligence. When physical science attempts to dogmatize over mind and over morality, we must bid it back to its place, by reason of the higher nobility of a Mind that knows itself, and that is capable of *Virtue.*

I have at last spoken the word that unlocks to us Man's spiritual dignity;—a being capable of Virtue can not be included within the category of Nature. Say, if you will, that the human intelligence is but a function of the brain, and is therefore a purely physical development; none dare affirm that moral action is due to physical causation; for to ascribe an action to physical control is to take away the morality of the action, to make right and wrong, good and evil, impossible as moral qualities. And what degradation of Man and of the universe so sad, so vile as that which would annihilate Virtue by denying to the soul the power of moral choice and its allegiance to moral law! "Two objects," said Kant, "fill my soul with an ever-increasing admiration and respect—above us the starry heaven, within us the moral law." But Pantheism effaces the moral law from the soul by denying to the soul that personality which is the essence of morality and of freedom; instead of divinizing Man it degrades him to a soulless atom in a godless universe; for the negation of personality makes God himself an incident of the universe, a germ of life and motion in process of development, an unfinished Deity struggling for expansion through physical motion, and "God, minus the world, an impossibility."

All that is grand and heroic in human life and history, Right, Duty, Justice, Liberty, Virtue,—all for which human souls have braved physical torture and mocked at death, all for which this Nation struggled four weary bloody years, with the life-blood of its bravest and best—is a delusion, a nonentity, if Man be under bonds to Nature without personality, will, consciousness of his own. Better, then, build altars to fire, flood, and fate, than memorial halls to heroes who fought for the fiction of Freedom and the Right. Nay, nay, there *is* a virtue, for there is a soul.

If some Comus, master of the rocks and woods, should tempt us with his " crystal glass "—distilled from Nature's laboratory—by which

> " The human countenance,
> The express resemblance of the gods, is changed
> Into some brutish form of wolf, or bear,
> Or ounce, or tiger, hog, or bearded goat;"

or should wave over us his sensuous wand, till

> " Nerves are all chained up in alabaster;"

and having thus "immancled" its "corporal rind " with his compelling arts, should attempt the freedom of the mind itself—though Reason, Motion, Will should fail us, *Virtue* could yet deliver from the spell;—Virtue the eternal bride of Heaven, Virtue that makes the soul like herself heavenly and immortal. She alone is free:

> " She can teach ye how to climb
> Higher than the sphery chime;
> Or if Virtue feeble were,
> Heaven itself would stoop to her."

To sum up all, if you demand proof that Man is a spiritual being, and therefore *super*-natural, I give you *Language*— not as Huxley would persuade us, the mere product of the nerve-force of certain muscles acting upon the structural peculiarity of the glottis, but language as a *psychical* creation, whose particles and connectives are links of intelligence holding speech together under the laws of mind—language, the very textile of the soul, of all Man's products the least physical or mechanical. I give you *Conscience*, discriminating right from wrong, and giving quality to actions from their motives.

I give you *Law*, dictated by reason, by justice, and addressed to the moral sentiments of the soul. I give you *Religion*, with its promptings of reverence, of worship, of duty and love, and its hope of immortality that conquers life's deadliest fears, and gives to its sorrows a supernal dignity.

Our total argument for Man's supremacy may be embodied in one concrete Man, a fitting type of the normal sovereign of Nature. Of comely person, of symmetrical brain, of opulent and varied powers; in point of physical development and of psychical aptitude and endowment one of the best products of Nature's handiwork; loving Nature as the poet loves her, as the artist loves her, obeying her precepts of health, studying her principles of order and of beauty, he nevertheless rose above Nature as a king, and swayed his scepter over wide realms of language, of thought, and of mind. Travel made him cosmopolitan in taste, sentiment, and culture, so that he disproved, in his own broad and genial humanity, all control of local material causes over a true-born human spirit; history made him a familiar citizen of Greece and of Rome, companion of their choicest schools; philosophy acquainted him with all that was great in human thought and profound and subtile in human reasoning; and thus absorbing into himself all countries, all nations, all ages, all thinkers and actors on the stage of life as the common property of a kingly soul, he made Nature tributary to his will in the service of Humanity. Beneath his wand cereals and grasses waved for the agriculturist, streams leaped to serve the manufacturer, and the most dry and dreary calculations of statistical laws became full of life and virtue for the merchant and the statesman. The stars lent their silvery music to the cadence of his tongue, and winds and seas moved in accompaniment to his words. And thus having mastered nature, earth, and time, rising in moral purity and nobleness to the fellowship of the immortals, he consecrated every power to rescue his country and mankind from the material and the vile, and watched and waited for the moral victory till "unseen hands shifted for him the curtains of the dawn."

And now shall Materialism mock that splendid history by linking it to the trilobite here and to the death-worm there? Nay, our hearts that discerned in him here the offspring of God, will yield him there only to that realm of spiritual powers, to that presence of the divine and the immortal which he has described as the consummation of the science of the soul, the true μετὰ τὰ φυσικὰ—" after nature, after time, after life, after death." Let Materialism stand rebuked before the splendid memory of

Edward Everett.

NOTE.—Professor Owen now advocates "Development" under the name of the *Derivative Hypothesis of Life and Species*, which is substantially as follows: Rejecting the principle of the direct or miraculous creation of species, he recognizes a natural law or secondary cause as operative in the production of species in orderly succession and progression—"of plants, or vertebrates, or other groups of organisms, such cause being the servant of predetermining intelligent will." The principle of *adaptation to purpose* he finds dominating that community of organization which all naturalists trace throughout the kingdom of Nature. For instance, he believes that the relation of the horse to the *Palæotherium* as its generalized type can be traced through transitional forms, such as the *Hipparion;* and "the succession in time accords with the gradational modifications by which *Palæotherium* is linked on to *Equus*." These modifications Owen ascribes to a law of derivation, originating in an intelligent purpose, and finds proof of this in the fitness of the organization of the horse and ass for the needs of mankind, and the coincidence of the origin of Ungulates having equine modifications of the perissodactyle structure with the period immediately preceding, or coincident with, the earliest evidence of the Human Race. "Of all the quadrupedal servants of Man none have proved of more value to him, in peace or war, than the horse; none have co-operated with the advanced races more influentially in Man's destined mastery over the earth and its lower denizens. In all the modifications of the old palæotherium type to this end, the horse has acquired nobler proportions and higher faculties, more strength, more speed, with amenability to bit. . . . . . I believe the horse to have been predestined and prepared for Man. . . . Assuming that *Palæotherium* did ultimately become *Equus*, I gain no conception of the operation of the effective force by personifying as 'Nature' the aggregate of beings which compose the universe, or the laws which govern these beings, by giving to my personification an attribute which can properly be predicated only of intelligence, and by saying, 'Nature has selected the mid-hoof and rejected the others.'"

We are now prepared to appreciate the distinction between Owen's views of the Origin of Species and those of Darwin, with which they have sometimes been confounded.

1. Professor Owen holds, with Darwin, that actual races are modifications of those ancient races which are exemplified by fossil remains.

2. Professor Owen holds with Darwin, that the changes recognizable in the earth's surface were not sudden and violent in their nature, and that the extinction of species was not cataclysmal but regulated. In his view, the discoveries of transitional forms,—as in the example of *Palæotherium* and *Equus*—have in a good measure supplied the link between the species held to have perished by cataclysms. The origin of species, therefore, is to be ascribed to a secondary cause, and not to the successively-repeated acts of direct creation.

Professor Owen protests wisely against invoking miraculous power to initiate every distinct species. No doubt a too constant appeal to miracle to account for the obscure, tends to cheapen our estimate of the supernatural. The Bible often uses natural means in conjunction with supernatural agency,—as the plagues of Egypt and the passage of the Red Sea; and it never introduces the supernatural to account for what can be explained by natural causes. In studying those phenomena of Nature whose causes are hidden from our view, we should imitate this wise discrimination of the Bible; since, as Owen says, "the miracle, by the very multiplication of its manifestations, becomes incredible—inconsistent with any worthy conception of an all-seeing, all-provident Omnipotence."

3. Professor Owen agrees, therefore, with Darwin in the theory of development to this extent, that he traces the origin of existing species to extinct species, through the operation of a secondary cause.

4. But Darwin defines this secondary cause as "Natural Selection," which is simply the law of the strongest prevailing in "the battle of life" with external circumstances. By the principle of Atavism, indeed, he recognizes in the indi-

viduals of a variety the capacity of recurring, under certain conditions, to a more pronounced expression of some particular feature or quality of the original type. His hypothesis is not atheistic, nor materialistic, for Darwin holds expressly to "the view of life, with its several powers, having been originally breathed by the Creator into a few forms or into one." But he makes the change of external circumstances the force that, by calling out certain elements or adaptations in favored individuals of a variety, educes by degrees the qualities that eventually give character to a new species. At this point Owen diverges from Darwin.

Professor Owen says indeed, as strongly as Darwin himself could state it, " I have been led to recognize species as exemplifying the continuous operation of natural law, or secondary cause ; and that, not only successively, but progressively ; from the first embodiment of the Vertebrate idea under its old Ichthyic vestment until it became arrayed in the glorious garb of the Human form." But while thus conceding a law of development, he refers this to the guiding intelligence of the Creator. "Species owe as little to the accidental concurrence of environing circumstances as Kosmos depends on a fortuitous concourse of atoms. A purposive route of development and change, of correlation and interdependence, manifesting intelligent Will, is as determinable in the succession of races as in the development and organization of the individual. Generations do not vary accidentally, in any and every direction ; but in preordained, definite, and correlated courses." This evidence of purpose in each change affecting species leads Professor Owen to the hypothesis of *Derivation*—"that every species changes in time, by virtue of inherent tendencies thereto." This hypothesis " sees among the effects of the innate tendency to change, irrespective of altered surrounding circumstances, a manifestation of creative power in the variety and beauty of the results ; and, in the ultimate forthcoming of a being susceptible of appreciating such beauty, evidence of the preordaining of such relation of power to the appreciation ; and it also recognizes a purpose in the defined and preordained course, due to innate capacity or power of change, by which nomogenously-created* protozoa have risen to the higher forms of plants and animals."

But in ascribing this power of evolution and progression to a secondary cause operating by some hidden principle of derivation, Professor Owen never loses sight of the personal Creator. He illustrates his hypothesis by magnetic phenomena, "which exemplify one of those subtile, interchangeable—may we not say immaterial?—modes of force which endows the metal with the power of attracting, selecting, and making to move a substance extraneous to itself. It is analogically conceivable that the same CAUSE which has endowed His world with power convertible into magnetic, electric, thermotic, and other forms or modes of Force, has also added the conditions of conversion into the *vital* mode."

These views of Professor Owen are condensed from the fortieth chapter of his "Anatomy of Vertebrates," which was also published separately in the *American Journal of Science* for January, 1869. They suggest some important reflections touching the main topic of this volume.

1. It is unwise and unfair to impute materialistic or skeptical opinions to physicists simply because they adhere to physical terms and methods in investigating and describing the phenomena of Nature, and refer all those phenomena to material causes. The most rigid Naturist may believe in an intelligent First Cause of the universe, and apart from his naturalism in Science may believe in the Bible as a revelation from God. This both Darwin and Owen profess to do; and the latter says, expressly, "My faith in a future life and the resurrection of the dead rests on the ground of their being parts of a divine revelation." Both these scientists only carry farther back in the succession of things the point of contact with that divine Will which was the original cause of all.

Even Herbert Spencer, who denies "the *absolute* commencement of organic life," says of the notion of "spontane-

---

* That is, primitive life-forms brought into existence by law, and not by a miracle; from νόμος, law, and γένω, root of γίγνομαι, to become, or come into being.

ous generation," as commonly understood,—"That creatures having *quite specific structures* are evolved in the course of a few hours, without antecedents calculated to determine their specific forms, is to me incredible. Not only the established truths of biology, but the established truths of science in general, negative the supposition that organisms, having structures definite enough to identify them as belonging to known genera and species, can be produced in the absence of germs derived from antecedent organisms of the same genera and species. . . . . . . The very conception of spontaneity is wholly incongruous with the conception of evolution. . . . . . . No form of evolution, inorganic or organic, can be spontaneous; but in every instance the antecedent forces must be adequate in their qualities, kinds, and distributions to work the observed effects. . . . . . . The supposed spontaneous generation habitually occurs in menstrua that contain either organic matter, or matter originally derived from organisms; and such organic matter, proceeding in all known cases from organisms of a higher kind, implies the pre-existence of such higher organisms. By what kind of logic, then, is it inferrible that organic life was initiated after a manner like that in which *Infusoria* are said to be now spontaneously generated? Where, before life commenced, were the superior organisms from which these lowest organisms obtained their organic matter?"* On the other hand, Mr. Spencer denies the necessity of a "first organism," and maintains that "organic matter was not produced all at once, but was reached through steps;" that "every kind of being is conceived as a product of modifications, wrought by insensible gradations on a pre-existing kind of being; and this holds as fully of the supposed commencement of organic life as of all subsequent developments of organic life." But he fails to account for the infinitesimal *protein*, or whatever was the primary molecule, and hence his view of the origin of organic life, by an interminable process of evolution, still leaves a place in the unmeasured past for the operation of a spiritual

---

* *Appleton's Journal*, No. 18, p. 563, and No. 19, p. 598.

Power "before life commenced." Whence came the molecules? and Who or What started the course of Evolution?

The history of Professor Owen's opinions illustrates the instability of scientific theories. Since the publication of his "Palæontology" he has openly shifted his ground upon the doctrine of specific creation by the intervention of miraculous power. His reasons for reversing his judgment upon this point appear plausible, but no more so than were his earlier arguments upon the other side. He differs now from Cuvier, because inductive research has brought him before an array of facts not known to his great teacher. But he differs equally from Darwin in the hypothesis by which he accounts for these same facts, and they both differ widely from other naturalists in respect of facts as well as of hypothesis. The treacherous ground for scientists is the *hypothesis*. Here, they leave the firm foundation of physical facts for the uncertainties of speculation, into which there enters more or less of the metaphysical element; and the temptation is strong to substitute conjecture for fact, or to piece out the line of facts by plausible conjectures, or to shape facts to the hypothesis. As Professor Owen himself testifies, "a favorite theory may render us blind to facts which are opposed to our prepossessions."

Hence, those who hold to the Bible in its integrity as a revelation from God need not be disturbed by a scientific hypothesis of to-day that seems to contradict the letter of the Scriptures. Twenty years may show the hypothesis to be untenable, or modify the facts of which it was constructed. It becomes physicists to be modest in the assertion of theories, especially in the sciences of physiology, archæology, and geology, where so much remains to be explored or revised; and it equally becomes biblicists to be modest in condemning a theory of Science upon the authority of the Bible, when there is yet so much to be learned in regard to the interpretation of the Scriptures. Experience has thus far shown that any true result in Science tends to harmonize with a true interpretation of the Bible.

3. For the right comprehension of the physical universe it

is necessary that we take into view spiritual powers as well as physical phenomena. The mere physicist can not include psychology within his proper province, yet there are facts of psychology that are as certain in consciousness as are physical facts to the senses; and as in the human organism the interaction of spirit and body is so complex and constant that no just idea of Man can be formed which does not include them both, so in the physical universe there may be invisible operations of spiritual powers which a higher Science of the spiritual would enable us to comprehend. Hence the denial of the supernatural may be a result, not of Science, but of ignorance. No one knows enough of the nature of spirit, or the modes of its operation, to be competent to affirm that a Spirit of infinite wisdom and power can not so act upon the material universe as to effect the purposes of an intelligent will, without seeming to disturb that course of things which, by reason of its apparent uniformity, we call the Laws of Nature. At the same time, the more refined those laws become in their working, the nearer do they bring one to the realm of spiritual powers.

LECTURE V.

The Antiquity of Man.

In following the narrative given in the 1st and 2d chapters of Genesis, we have seen that the creation of Man was ushered in by a distinctive formula differing very significantly from that which describes the preceding periods of creation, and denoting some more exalted purpose on the part of the Creator. Among all the forms of organized being hitherto produced, there was no type for this intended lord of the creation, and accordingly he was made directly in the image and likeness of God, intellectually and spiritually, as a moral being, and in this character of sonship was established at once in dominion over all the other works of God in this lower world. Your attention was called to the fact that Nature—by which we mean nothing independent of God, but that course of things in continuity which is established as the result of the creative power—was in existence before Man, with its established elements, principles, forces, and laws, as we now know them (a fact which is very clear upon the face of the geological records), and also that there were immense æons in the process of this earth's formation during which Man could not possibly have existed in its atmospheric and other conditions. We have further seen that there was nothing in those pre-existing forces and elements of Nature capable of producing Man; that while there has been a law of advance in type-forms through the whole course of creation, there has been also a lifting up by a power coming in from above upon antecedent conditions, which conditions, though necessary to the next step, could not of themselves create or originate that step. We have seen, also, that the notion of a development of the human species from any preceding form of organization has nothing to substantiate it in anything yet discovered, and is unphilosophical in the assumption with which it starts. We have seen, moreover, that Man exists not in and of Nature, as a mere part of it, but over Nature, as ruling it and subjecting it to his uses. We should properly have closed the consideration of the *status* of Man at that point, had it not been that the question of his Antiquity, though by no means settled upon valid grounds of Science, has assumed an importance that demands the recognition of all fair-minded theologians.

There is a great deal of superficial talk nowadays concerning the alleged discoveries and determinations of Science upon this question, as being both in fact and in tone hostile to the Scriptures. Some scientists, indeed, parade their hostility to the Bible, and some theologians are as forward in their denunciations of Science; but there is no just ground for such opposition upon either side.

Those who have been accustomed to hear discourses from this pulpit, will bear witness that there is no philosophical lecture-room in the land where true Science is more uniformly accredited and held in honor than here; that no fact of Science is ever questioned, no established principle of Science is ever gainsaid, but all that Science has really proved is gratefully accepted as part of God's teachings to Mankind, in the conviction that there never has been and never can be any real collision between facts of Science fairly made out and the Bible fairly interpreted. In this spirit we propose to take up the question of the Antiquity of Man, to bring out fairly all the facts that bear upon it, so far as they are known, and to lay these side by side with the record of the Word of God.

How long has Man *been upon the globe?* I do not know.

Does anybody know? Are we able to trace back the human race to its beginning, and to measure the term of its duration? Not yet, I think. The data upon this subject are meagre and uncertain, and the question, which ought to be simply one of fact, resolves itself pretty much into one of speculative or problematical inquiry. Hence, when we study it purely as a question in Natural History, we should keep distinctly in our minds the only fact that as yet is a fact about it, viz., that it is extremely problematical. Setting aside for the present this narrative in Genesis, what data have we by which to determine the continuance of Man upon the globe?

1. There are *monumental remains* scattered here and there upon the surface of the globe which are supposed to belong to a remote antiquity. But these fall within *measurable* periods of time. There may be here and there a mooted question as to the age of some particular monuments of antiquity; but none of these make very extravagant demands upon our faith. Take, for instance, the probable age of the pyramids of Egypt. I say *probable*, since it is not yet possible from the sources at command to fix precisely the date of their erection in the chronology of the world, although their place in the dynasties of Egypt is more nearly ascertained. The great pyramids, by the common consent of Egyptologers, are assigned to the Fourth dynasty of kings of the Old Empire, as given by Manetho; and the commencement of this dynasty has been placed by Lepsius, Bunsen, Brugsch, and others, at from three thousand to three thousand six hundred years before Christ. Bunsen, however, in his latest recension of Egyptian chronology, revised Manetho's lists by those of Eratosthenes of Alexandria, and thus, as he expresses it, "got rid legitimately of a considerable number of useless centuries." He had before said, "In no part of *Asia* does chronological national history go back beyond

4000 B.C., though we see everywhere traces of a preceding epoch of tribes and municipal cities;" and now in *Egypt* he reduces the epoch of Menes to 3059 B.C., and the date of the three great pyramids from 2645 B.C. to 2559 B.C.,* an abatement of six hundred years from his previous estimate. Mr. C. Piazzi Smith, who looks upon the Great Pyramid as a monument of mathematical and astronomical science, professes to have determined from astronomical data the erection of the Great Pyramid as in the year 2170 B.C.; much of his reasoning, however, is theoretical. But if we take the extremest view of responsible authorities in Egyptology, these do not attempt to place the pyramids farther back than four thousand years before Christ; and this, the oldest conjectural period, is purely *conjectural* on the part of those who advocate the longest chronology for the Egyptian empire. Even should we concede such antiquity to these monuments, this is still *measurable*, for these are periods of time that we can comprehend. We could adjust our chronology to such dates, without taxing credulity overmuch in respect to the main question of the Antiquity of Man.

2. Leaving, then, these monuments upon an appreciable historic basis, we pass in the next place to the *traditions* of a remote origin, which are found among various nations. With the perhaps solitary exception of the Hebrews, these traditions reach back to a far antiquity. But we detect in them all a fabulous element. The moment one passes over the sharp line of definite history, one is in a dim and shadowy land, where there is nothing authentic to rest upon or work with. Two views may be taken of these traditions. On the one side it is claimed, with a show of reason, that there must have been some basis at least for the origin of national traditions

---

* "Egypt's Place," vol. v., p. 62.

of a vague antiquity—some prolonged indefinite period before we reach the positive beginnings of history. But, on the other hand, it may be asserted with equal significance, that there is a tendency in the human mind universally to invent and exaggerate matters pertaining to the remote past. Even within a period that barely transcends the memory of the living, how many stories we have accumulated already about Washington, magnifying his heroic character and life, his remarkable preservation from danger, and so on—some of which probably have not a shadow of foundation in fact. How many stories we have about the Pilgrim Fathers, and the early history of New England, which probably have no warrant whatever in fact, and yet these go back only two centuries. When we go back ten centuries, how much of tradition must needs be mythical; and when it comes to periods of thousands of years, the whole thing is so indeterminate that no philosophical mind can accept it. Hence, judicious investigators of Roman and Grecian history—Niebuhr, Mommsen, Grote—have sifted out and thrown aside as worthless a great mass of early tradition. These two data, Monuments and Traditions, are of comparatively little help in determining with accuracy the period of Man's beginning, the term of his continuance upon this globe.

3. We pass, then, in the third place, to certain *remains of human workmanship* which are found in such relations to extinct races of animals, or in such geological conditions as mark a high antiquity. Let me enumerate a few of these. In 1853–4, during a remarkably dry winter in Switzerland, the lakes and rivers fell far below their usual level, and the inhabitants of Meilen, on the Lake of Zurich, improved the opportunity in dredging and building walls, and in so doing came upon the remains of piles and various traces of human habitations. Explorations were made in the lakes generally, and

in Zurich, Constance, Geneva, Neufchatel, Bienne, Morat, Sempach, and in many of the smaller lakes, Inkwyl, Pfeflikon, Moosseedorf, Luissel, etc., were found remains of pile-habitations, numbering in the whole more than two hundred distinct settlements. These were remains of piles and of platforms upon which houses had been built, remains of weapons and of domestic utensils, the bones of animals, and now and then bones of men. But these diversified remains were so confusedly mingled that it was impossible to assign them all to any one era, or in some cases to fix definitely upon any period for the construction of the dwellings. Some of them clearly belonged to Roman times; others to a period when iron had come into general use; while others exhibited only traces of bronze or of stone; but as yet, in these lake settlements, nothing has been found that would necessarily carry us back to a remarkable antiquity. Indeed, the remains of domestic animals and of the cereals found in these lake settlements bring them within the pale of a comparative civilization. Herodotus describes such lake-dwellers. A principal point of interest and difficulty in connection with these lake habitations is the occurrence of peat, which in some instances has been found deposited to a considerable depth over the remains of human settlements. It is assumed by some that the rate of formation of peat is so slow as to require a very long period for the growth of the peat-bog at Chamblon, on the Lake of Neufchatel, or the morass at Pont de Thièle, in the Lake of Bienne. That, however, is a question in dispute among scientific men; and we must wait until there is some satisfactory evidence as to the length of time required for such peat deposits as those just mentioned, before we attempt to pronounce any definite opinion. True science is always modest, —waits until it knows. The remark of Sir Charles Lyell is here in point: "The depth of overlying peat affords no safe criterion for calculating the age of the cabin or village, for both in England and Ireland, within historical times, bogs have burst and sent forth great volumes of black mud, which has been known to creep over the country at a slow pace, flowing somewhat at the rate of ordinary lava-currents, and sometimes overwhelming woods and cottages, and leaving a deposit upon them of bog-earth fifteen feet thick."* To the same effect is the admission of Carl Vogt, that "we neither know generally within what time a stratum of peat one foot thick may grow, nor do we possess any scientific data to calculate the quantity of growth within a given time of any individual peat-moor; that the growth must differ in various moors; that even in a given locality it must have differed during certain periods, is easily imaginable." †

In the investigation of a problem so obscure as this of the Antiquity of Man, where the temptation is strong to make hasty generalizations and unfounded assertions, it is well to keep in mind the famous maxim of Confucius, that "Knowledge consists in *knowing* what we know, and also in knowing *what* we do not know ;" and the knowledge of our ignorance, the ability to make a sharp distinction between the known and the unknown, will save us from many false conclusions.

4. As the argument for the Antiquity of Man is cumulative, and I desire to give to each item its proportionate weight, I pass from the lake-dwellings to the evidence from *mounds* of various descriptions in different parts of the globe. Those on the coasts of Denmark, known as "Kitchen-Middens," are probably made up of the refuse deposited from kitchens by the aboriginal hunters and fishers of those islands. These mounds contain shells, bones, and utensils. The bones are exclusively

---

* "Antiquity of Man," p. 32.
† Art. in "*Archiv. für Anthropologie*," translated in *Anthropological Review*, No. 17, p. 209.

those of animals that have inhabited Europe within historic times; the shells are of living species, but no longer found in the waters adjacent to the mounds; the implements are of stone. Accordingly there is nothing in these findings that points to a very high antiquity.

The pretentious claim of a vast antiquity for corresponding mounds in Scotland, put forth by Mr. Laing in his "Pre-Historic Remains of Caithness," has been exposed by J. Cleghorn, Esq., a scientific antiquary, who shows conclusively, upon geological and archæological grounds, that remains which Mr. Laing ascribed to unknown races in pre-historic times, can not date back more than three or four hundred years.* The mounds are thus far a very indeterminate element in any calculation of the Antiquity of Man.

Traces of human works and habitations have also been found in Denmark, in peat-bogs in connection with remains of pine and oak forests, which have been long supplanted by forests of beech, which covered the Danish isles as far back as the time of the Romans. Much time must be allowed for such a succession of forests, and for so thick a deposit of peat. But here, as before, all estimates are as yet conjectural, and no data yet supplied from these sources demand the immense periods that some assign to Man's existence upon the globe. "Differences in the humidity of the climate, or in the intensity and duration of summer's heat and winter's cold, as well as diversity in the species of plants which most abound, would cause the peat to grow more or less rapidly, not only when we compare two distinct countries in Europe, but the same country at two successive periods." †

---

* See this whole question discussed in the *Anthropological Review*, and Journal of the Anthropological Society, No. 14, p. cxxxix. See also the masterly work of Nilsson, "*Les Habitants primitifs* de la Scandinavie," one of the best authorities on the subject.

† Sir Charles Lyell, "Antiquity of Man," p. 111.

A more important point, and one that throws us farther back in the uncertain period of Man's origin, is the *remains found in the caves of rivers*—as for instance in Belgium, where they have been minutely explored—or in the sides of rocks, where rivers once had their bed. Here are implements evidently fashioned by the hands of men, and along with them the bones of extinct species of animals, such as the cave-bear hyena, elephant, and rhinoceros. Similar remains have been found in France, especially in the valleys of the Seine and the Somme, in the deposits known as the river-drift, consisting of strata of sandy marl, gravel, etc. Where such layers are found at a great elevation above the present level of the stream, in addition to the computed age of the drift, time must be allowed for the working down of so enormous a mass as the stream has bored through in reaching its present bed. How much time can not be determined, for it is yet an unsettled question in Geology, whether certain agencies concerned in such changes did not work more rapidly in the earlier formative periods of our globe than they are seen or computed to work at present, or whether convulsions may not sometimes have precipitated what is assumed to have been the result of gradual changes. Moreover, "alluvial formations still take place in which the products of former and later periods are commingled. Thus a river running through sand-banks belonging to different periods of formation may mingle portions of these sand-banks in a new alluvial formation. . . . . Deposits may take place which, coming from heights, may present features of being older than they actually are. It is exceedingly difficult to parallelize the various deposits characterizing the diluvium, and for the present, at least, to determine the chronological succession in which certain formations of different countries stand to each other, especially as the stratification which guides us in elder beds is in the diluvium very confused, and *offers no certain basis for tenable inferences.** But there is no room to question the general result of these researches among the river-caves and the diluvial drift; the findings are too numerous and well attested, and the archæological and geological conditions too well ascertained, to admit a doubt that Man existed in Europe contemporaneously with the cave-bear, and at least upon the margin of the glacial age. What, then, shall we make of these facts in view of the Biblical account of the origin of Man?

Although the question of the Antiquity of Man is by no means a novel one, this phase of it is so new that as yet no one is in a position to pronounce upon it with final authority. There are not data enough for absolute conclusions. Two tendencies, however, should be guarded against: first, the speculative tendency in the minds of some who are exclusively devoted to physical science, to assume a great age for every newly discovered fact on the face of the globe. This tendency to theorize is not common with scientific men; yet there are men of true science who have the speculative tendency very strongly developed. In reading the writings of such physicists, one should be on his guard against this continually, and watch them closely where, perhaps unconsciously to themselves, they drop out of the line of facts into the line of theory and speculation. Especially should one receive with caution such facts as first appear in the newspapers! Many of you will remember the sensation caused, a few years ago, by the announcement that there had been brought up from a depth of some ninety feet or more under the alluvial deposits of the Nile a piece of pottery indicating human workmanship. And then there were profound calculations to show how many thousands of years old this

---

* Carl Vogt, *Anthropological Review*, No. 17, p. 208.

deposit was—measuring by the rate of formation in the Nile delta—and finally this was placed at a figure so enormous that one was staggered at the attempt to conceive of Man as having existed for billions of years. This "discovery" went through the newspapers, "and where is Moses? where is Genesis?" was the cry; until a more careful investigation proved to the satisfaction of experienced archæologists that the bit of pottery was of *Roman* origin, which ended that matter. Some will remember the great hurrah that went up from the camp of infidelity when the zodiac on the temple of Denderah was first discovered, and what an immense stretch of time was assigned to the structure of that edifice as marked by this astronomical ornament! But how did the reading of the hieroglyphics shame the rashness of such antiquaries! Hence, it is well to guard against the purely speculative habit of ascribing an immense age to every new discovery in archæology.

At the same time, I would earnestly exhort theologians, and all Christians, to guard against the tendency on the other side—to raise the cry of infidelity or skepticism against men of Science for every theory that they propound which is not in obvious harmony with the Bible. That is not the way to deal with these questions on either side. I make no pretension to being a man of Science; but as an interpreter of the Bible, I am as much beholden to any fact of Science as the most accomplished scientist. We are not warranted in pitting Science and the Scriptures one against the other. It is not philosophical in the man of Science to raise a hue and cry against the Bible as soon as he discovers something new; but, on the other hand, let the theologian be careful how he raises the cry of infidelity. Now, there are certain broad principles that we can lay down which may govern us in the further investigation of the question of Man's antiquity.

DID THE HUMAN RACE BEGIN IN BARBARISM? 95

1. Whatever may prove to be the fact in regard to the Antiquity of Man, it is a groundless assumption that Man began his existence at a low stage of barbarism.

Some have projected a theory of the beginning and growth of human society to the following effect:—that Man began at an indefinitely remote period, and at the lowest conceivable stage, using stone only for his implements;—this was the "stone age;" by and by he advanced a little, and came to the discovery and use of bronze—hence the "bronze age;" after that came the use of iron, and so the "iron age;" and these successive gradations are supposed to mark the origin and progress of the race. But it is sheer assumption that this indicates the universal history of Mankind.

Stone implements have been found here and there in the lakes, caves, mounds, and river-drift of Europe. What then? The men who made those implements may have possessed intelligence and a moral development far above what these manifestations would indicate, while as yet they had only an imperfect knowledge of the arts. The Hebrew patriarchs, of a much later time, were not advanced in the industrial arts in proportion to their moral development. Moreover, great ingenuity and resource are shown by many of the rudest tribes in their weapons, and the sense of beauty is evinced by them in the choice and invention of ornamental forms. It is a hasty conclusion to assume that Mankind everywhere began their existence at a very low stage of barbarism, simply because we have found in certain localities only an inferior kind of implements and weapons.

According to Mr. Bickmore, in the *Ki* group of the East Indian Archipelago, "the natives are very industrious and famous as boat-builders. They *need no iron* to complete boats of considerable size, which they sell to inhabitants of all that part of the archipelago." Here are mechanics and traders living without iron in the very millennium of the "iron age" of the world.

2. It is also a groundless assumption that if the "stone age," so called, existed, it was universal at any one time upon the habitable globe. This "stone age" exists to-day among certain savage tribes, and is contemporary with the steam engine, the locomotive, the magnetic telegraph, that are types and indications of man's highest civilization in material things.

Look at the condition of that Britain, from whose loins our ancestors sprang, at the time when Constantinople was the seat of a luxurious court. "Her shores were, to the polished race which dwelt by the Bosporus, objects of a mysterious horror, such as that with which the Ionians of the age of Homer had regarded the Straits of Scylla and the city of the Læstrygonian cannibals."* Or go back a little earlier, to the period when the Romans first invaded Britain; history tells us that the conquerors even despaired of teaching such barbarians how to build a stone wall. There was the luxurious civilization of Italy and the East, side by side with this stone age in Western Europe. Consider how small a portion of the globe has yet been examined for the relics of antiquity. The vast fields of Central Asia are probably rich in deposits of the earlier periods of humanity; and these have been but little explored.

Carl Vogt admits that the Stone, Bronze, and Iron periods may overlap each other. "The Homeric heroes, who knew bronze and iron, threw stones at each other, and the sling was, in not very remote times, a legitimate war weapon. Certain as it is that stone, bronze, and iron periods only form relative sections continued into one another, it can not be

---

* Macaulay, "History of England," chap. 1.

assumed that similar civilization epochs were simultaneously developed in different parts of the globe. In other words, even in the limited area of Europe, there may, on the coasts and on rivers, peoples have existed, further advanced in civilization, who knew of and how to use metals; while in the interior of the country tribes dwelt, who for centuries, perhaps, had no idea of metals, not unlike the savages of islands who used stone weapons until Europe supplied them with iron, lead, and powder. . . . . As in our present civilization epoch there are many regions where Man requires his whole time for the acquirement of the necessities of existence, so must it in greater degree have been in primitive times; and thus it might have come to pass, that while in one district civilization had sufficiently progressed for the manufacture of more perfect implements, those of adjoining districts were still in a rudimentary condition."* It is well to be modest in asserting the universality of a stone age, contemporary all over the globe.

3. It is a groundless assumption that the stone age was the first type of human existence anywhere. It may have marked deterioration. What is there, especially until Central Asia shall have been explored, to disprove the representation made by the first chapter of Genesis, that man began his existence fitted by his Creator for the work of subjugating Nature, and began at once to do this? But with the progress of human society, with the increase of population, what is the tendency? Always to crowd the weaker off to the farthest outposts. And what is the consequence of that? Deterioration always, unless the stronger find some motive, as for instance in the discovery of gold or in fat fields inviting them to go after the weaker, and carry with them their

---

* *Anthropological Review*, No. 17, pp. 205, 214.

civilization. The lost arts are an indication of what may often have repeated itself in human history. "Flint implements are a very uncertain indication of civilization even among the tribes who used them; and are no index at all of the state of civilization among other tribes who lived at the same time in other portions of the globe."*

4. It is a theoretical assumption—not a demonstrated fact—that the present rate of geological changes, as for instance in the formation of deltas or the uplifting and depression of continents, is the proper guage for measuring such changes in the past. Some very eminent geologists are of opinion that " the causes of geological change were once more intense and rapid in their action than they are now." One such authority, writing in the *Anthropological Review*, makes these concessions: "It may almost be asserted that every scientific opinion is speculative. It may be safely said that there is no opinion current among scientific men,—not even of those opinions whose claim to the title 'principle' appears most unquestionable,—that is not essentially *provisional*, liable to modification or even revolution under the pressure of increased knowledge;" and applying this to theories of the origin of Man, he speaks of the "present incomplete state of our knowledge, and the necessity of waiting for a larger and clearer mass of testimony before venturing to try conclusions upon a subject so obscure."† The several considerations here enumerated show how premature and unauthoritative would be a judgment just now in favor of the extreme antiquity of Man upon the globe.

But, on the other hand, there are facts that seem to call for an extension of time considerably beyond the computed chronology of the Bible, in order to admit of all that has been

---

* Duke of Argyll, "Primeval Man." † *Anthropological Review*, No. 24, p. 19.

effected by Man and in Man since his first appearance on the earth.

1. The oldest monuments of Egypt can hardly be brought within the date of the flood of Noah according to the received Hebrew chronology. The date assigned to the three great pyramids by most Egyptologists is older than the flood as this is reckoned in the margin of our Bibles; and the lowest date to which Prof. Piazzi Smith and other advocates of the shorter chronology would reduce them by astronomical modes of computation, is still far too old to be fairly accommodated to the Hebrew date of the flood; for the building of those stupendous monuments required a knowledge of the mechanic arts that perplexes us even in this day of mechanical inventions, and a consolidated state of society under a centralized power that only time could have established. Such a condition of things is not reached in a day. I have shown, indeed, how groundless is the assumption that Man began his existence at the low stage of barbarism of which flint implements are taken to be the index; and that there is no warrant of fact or of philosophy for the thousands of years laid down by Bunsen as necessary for the development of a condition of civilization equal to the building of the pyramids. On the contrary, the theory is quite as plausible that Man began his existence under mental and material conditions that favored the rapid construction of a civilized society, and that the remains of a primitive barbarism are also tokens of deterioration from the original type of Humanity. It is plain from the Hebrew narrative that, at the time of the flood, the mechanic arts were in a good state of forwardness. As far back as the days of Lamech we read of artificers in brass and iron, the invention of musical instruments, the building of cities; and surely so huge a craft as the ark, with its interior stories and divisions, required no mean skill for its construction.

The knowledge of the arts must of course have been transmitted from antediluvian times through the survivors of the flood; and we read, soon after, of the building of great cities, and of the tower of Babel. But for such works, and especially for founding such an empire as was ancient Egypt, there was need of centuries for the growth of a population in numbers and resources equal to the gigantic structures that crown the banks of the Nile. The less than two centuries between Archbishop Usher's date of the cessation of the flood and Piazzi Smith's calculation of the date of the Great Pyramid, was far too short an interval for results upon a scale so magnificent.

The Tablet of Sethos I., recently discovered in the great temple of Abydos, introduces a new element of complication into these calculations. Upon this Tablet a monarch whose period is pretty clearly determined as of the fifteenth century before Christ, is represented as offering sacrifice to his royal predecessors, of whom there are seventy-six in an unbroken line up to Menes; and this line tallies with the fragmentary lists from other sources, showing that this was the official list of recognized sovereigns in regular succession. Eight reigns in a century would by the analogy of history, in long periods, be a large allowance. This is greater than the average for the thousand years of English history from Egbert to Victoria, through all the changes of Anglo-Saxons, Danes, Saxons, Normans, the Plantagenets, the contests of Lancaster and York, the Tudors, the Stuarts, the Commonwealth, and the Revolution. In settled times the average is not over five to a century. But even an average of ten reigns in a century would require the whole time from Sethos I. back to the Flood of our common chronology to dispose of the seventy-six predecessors of that king. And when we have arrived at Menes, we find already an empire consolidated from previous district governments, and capable of building the great city of Memphis, with its magnificent temples and towers, and its huge dyke that turned the course of the Nile. Either, then, we must place the Flood much farther back upon the chronological scale, or must admit not only that it was not universal in territorial extent (which is altogether probable), but that it was not universal in the destruction of mankind, which would seem to contradict both the letter and the spirit of the sacred record.

2. The unchanged appearance of leading types of mankind, as far back as we can trace these in history, requires a considerable extension of time to account for their origin, provided we adhere to the physiological unity of the race. Upon Egyptian monuments that date back from 1400 to 2000 years before Christ, the negro is depicted with color and features as marked and characteristic as he exhibits at this day. When did this type originate, which has remained unchanged for more than three thousand years? If the type itself was a gradual product of time, how *much* time before the date when it begins to appear upon Egyptian monuments was necessary to establish its marked and unvarying features? According to a tablet of Sethos I. the Egyptians divided mankind into four principal races—the Red (Egyptian), the Yellow (Ammonites), the Black (Negroes), and the White (Libyans). If all mankind were descended from a single pair; and again, if the whole peopled earth was destroyed by the flood, with the solitary exception of the family of Noah, how much time was required to originate peculiarities of race which can be traced back without variation through the whole known course of history? In the present state of scientific knowledge, this whole subject is wrapped in obscurity.

3. The formation of Language, and its distribution into the

great classes of human speech, call for an extension of time, if one adheres to the belief that all languages were derived from one primitive root, which is only another form of the doctrine of the unity of the race. The lines of language converge toward Central Asia, and in the far past its many threads can be woven into a small number of strands, which the science of Comparative Philology may yet succeed in twisting together in a single cord; but cautious philologists doubt whether conclusive testimony for or against the unity of the human race will ever be derived from language alone. If there was one primitive language of the race, the Biblical story of the confusion of tongues at Babel would account for the diversities of human speech. But when the trustworthiness of the Biblical narrative is under consideration, we have no right to assume the miraculous element as a mode of meeting difficulties that seem to embarrass the narrative itself. That it is difficult to provide for a normal division of tongues from one primitive root within the period of our received Chronology, must be obvious to any who will reflect upon the elements that enter into the construction and growth of language.

4. Man in the fossil state, although rarely found, is another element of perplexity in the question of his antiquity. To be sure, it does not require any great length of time, nowadays, to produce a human fossil! There were many such in the churches and in politics in the old times of slavery! Nor is the time required for producing a proper fossil so great as the popular mind is apt to imagine. To persons unfamiliar with geology, human skeletons in a fossil state are a great wonder; but some of the most complete of these are of recent origin. For instance, such skeletons found in the island of Guadaloupe—one of which is in the British Museum—were taken from a shell limestone of modern origin, and which is still in

MAN AT THE CLOSE OF THE GLACIAL PERIOD. 103

process of formation. These are the remains of Caribs killed in battle not over two hundred years ago.*

Professor Guyot,† and others of the more conservative school of geologists, are disposed to admit the existence of Man during the glacial age. There are traces in the Northern Hemisphere of two great ice-periods, each succeeded by a time of quiet, and this again by fluvial action. The same periods were marked by great oscillations, and especially by the submergence of the land northward. That life coexisted with this condition of the globe, is evident from the remains of the elephant, the rhinoceros, the hippopotamus, etc.; and the evidences of Man's existence during the later glacial period are to be found just where one should look for them—in the terraces along the hills, in the peat bogs in the valleys, and in the river caves—localities in which human implements and bones have been discovered. But Professor Guyot calculates, from astronomical data, that this great ice age was not more than nine or ten thousand years back. This, however, is a period that defies our present chronological scales, and we can only wait for further light. Prestwich, a candid and able investigator, makes a suggestion of some weight for abbreviating this relative antiquity: "The evidence from the occurrence of human relics, with the bones of extinct animals, as it at present stands, does not seem to me to necessitate the carrying of Man back in past time, so much as the bringing forward the extinct animals toward our own time."

But though at present we must despair of any definite conclusions upon the Antiquity of Man, there are certain principles of adjustment between Science and the Scriptures which we should hold distinctly in view.

---

* Dana's "Manual of Geology," p. 580.   † The "Morse" Lectures for 1869.

Chronology is not minutely mapped out in the Bible. The order of succession is given without reference always to the scale of time; the facts themselves, as stated in the book of Genesis, may all be true; the succession of events there recorded may be also true and correct; the creation, the temptation and fall, the dispersion, the flood, the after-dispersion and migration of the nations—these all may be true as facts, and yet we may not be able to adjust to them properly a sliding scale of Chronology in respect to years.

The Duke of Argyll defines as the Chronology of "Time-relative," time which is measurable, not by years, but only by an ascertained order or succession of events. After we pass the limits of known history, this Chronology of *order* is the only form in which events can be measured. In pre-historic times, and also in the shadowy twilight where tradition begins to assume the shape of history, Chronology is necessarily obscure—indeed, one of the most obscure topics that science and history have to deal with. It is particularly obscure and difficult when we have to do with Oriental modes of computation, which are essentially different from ours. Before the time of Abraham, the narrative given in the Book of Genesis may be a condensed epitome of foregoing history—not a consecutive line of historical events year by year, and generation by generation, but a condensed epitome of what had occurred in the world from the creation to that time; for if you will scrutinize it carefully, you will see that in some instances the names of individuals are put for tribes, dynasties, and nations, and that it is no part of the object of the historian to give the consecutive course of affairs in the world at large. I have much hope in this matter from the translation of the Bible into the Arabic, and its wide diffusion among a people cognate to the Hebrews. I have no doubt that there is yet to come to us from Arabian and other

Oriental sources a mode of interpreting Chronology according to these lists of names, which I do not believe we have yet fairly got hold of, and therefore I am not troubled by any seeming discrepancies. New light may arise upon the science of interpretation; and when we shall have learned more about the idioms of Oriental nations in matters of Chronology, we may obtain all the margin that we shall require in the matter of Man's antiquity—that is, so far as monuments and traditions are concerned. Moreover, the early history may be, in some sense, typical—describing the *typical* Man as God made him to be, and how he was placed in respect to Nature before he sinned. Yet, in this very matter of Chronology, there are internal evidences of truth in this early Hebrew narration, as contrasted with other ancient stories, in these two principles: *First*, the absence of those immense vague periods which precede the historical chronologies of all other people; the absence of that legendary phase of things which is so marked in the history of the Egyptians and Hindoos. And *secondly*, there is no attempt in this early history to magnify the Jewish people. Every other nation of antiquity sets out to magnify itself as descended from the gods—perhaps as having originated upon the soil it occupies, and, being thus favored of Heaven, as having come down from an immense antiquity.

Now, there is no such endeavor to magnify the Jewish people in this Biblical story. It does not place the beginning of Man in the country destined to be their country, but off in the far East; and it gives the history of Man as MAN, and with no attempt at self-glorification. All this is significant of the historical in contrast with the mythical style. At the same time this history has ever in view one definite purpose; it does not profess to be a secular history of Mankind, but the history of certain typical or elect persons, families, and races, given to illustrate the providential and moral government of God. Thus, after the fall of Adam and the crime of Cain, *Seth* is the selected typical Man, and the course of the history runs mainly in the line of his posterity; and when in the course of ages the descendants of Seth apostatize, *Noah*, who remains faithful, is selected as the typical Man, and is carried through the flood to become the founder of a new world. In like manner, when the descendants of Noah have become degenerate, *Abraham* is selected to be the Father of the Faithful, the head of a commonwealth both civil and spiritual that shall thereafter be in the world as the kingdom of God. Now, some would apply this obvious principle of selection in the early Biblical history to the case of Adam, and regard him, not as strictly the first man created and the sole progenitor of the human race, but the first called to a representative position as the Son of God, and the head of a new type of Humanity. Such was the doctrine of Perriere, of Bordeaux, in the seventeenth century, in his famous treatise on the Pre-Adamites,* which has been revived of late by an anonymous English author.† Some plausible arguments are urged for this opinion; for instance, that Cain, in going forth from the home of Adam, expected to find the regions round about peopled, and that their inhabitants would abhor him because of his crime; that he must have found a wife among other tribes, or have married his own sister, contrary to the divine law of marriage; that he built a city, for which he must have had both laborers and a population. It is further argued that, in the first chapter of Genesis, Man is introduced as the close of the geological system, and is there spoken of in a general way; and that, after a long interval,

---

* "*Præ-Adamitæ*, sive Exercitatio super versibus 12, 13, and 14, Cap. V. Epistolæ D. Pauli ad Romanos. Quibus indicuntur primi Homines ante Adamum conditi."

† The "Genesis of the Earth and of Man," Edinburgh, 1856.

the second chapter of Genesis introduces a particular type of Man, the *Adamite*, as representing the beginning of history under a special divine plan. After the world had long been peopled with rude barbarian races, this diviner Man was introduced with a view to a kingdom of spiritual power; but his apostasy frustrated for a time the divine purpose. Nevertheless in the seed of Seth the worshipers of the true God were for a time revived; but when these "sons of God" became enamored of the daughters of the old sensuous, idolatrous stock around them, and entered into an unholy alliance, the flood was sent to exterminate that apostate seed. Such is the theory, and although open to some serious objections, it serves to show one *possible* way in which the Bible and Science may yet be harmonized upon the question of the Antiquity of Man and the unity of the race. It may prove eventually that there is in this brief record in Genesis a margin for all the discoveries of Science.

For the present, however, we must accept whatever is clearly established as *fact*, even though we can not fully reconcile one class of facts with another. It is unscientific to frame a theory upon one class of facts in opposition to other facts that rest upon reasonable evidence. Whatever conclusion may at last be reached concerning the Antiquity of Man, there are two truths established in the foregoing lectures that can not be affected by it.

1. The *succession of events* in the creation is unimpaired, and Man appeared in the order assigned to him as the latest and highest work of the Creator. "It is not known that any new species of plants or animals have appeared on the earth since the creation of Man."* He was the crown of the whole series.

---

* Prof. Dana.

2. Whenever Man began to be, he had, at least in rudimentary exercise, the place of dominion over Nature to which the Bible appoints him. "Even in the most rudimentary form, the use of an implement fashioned for a special purpose is absolutely peculiar to Man. There is quite as much ingenuity and skill in the manufacture of a knife of flint as in the manufacture of a knife of iron. As regards his characteristic mental powers, Man has always been Man, and nothing less."* That able scientist, so often quoted, Carl Vogt, also admits the strong human characteristics of Man in the most primitive period. "This powerful, tall, and strong primitive man, who lived by the side of the cave-bear and the mammoth, already honored his dead by burying them in grottoes closed with slabs, and furnished them with meat and arms for their journey into another world. He knew the use of fire, and constructed hearths, where he roasted his meat; for of pottery the traces are but few. His implements or weapons consist of rude hatchets and knives, which were struck off from a flint block by another stone, and of worked bones employed for handles, arrows, clubs, or awls. This wild, primitive man endeavored to ornament his person with perforated pieces of coral and the teeth of wild animals. He probably dressed in skins or prepared bark of trees."

There was Man at his starting-point, at an immeasurable distance in characteristic adaptations above any creature before him. We may fitly close this whole review of his origin and antiquity with these eloquent words of Professor Dana:

"When Man appears, the animal element is no longer dominant, but *Mind* in the possession of a being at the head of the kingdom of life. Man was the first being that was not

---

* Duke of Argyll, "Primeval Man."

finished on reaching adult growth, but was provided with powers for indefinite expansion, a will for a life of work, and boundless aspirations to lead to endless improvement. He was the first being capable of an intelligent survey of Nature and comprehension of her laws; the first capable of augmenting his strength by bending Nature to his service, rendering thereby a weak body stronger than all possible animal force; the first capable of deriving happiness from beauty, truth, and goodness; of apprehending eternal right; of looking from the finite toward the infinite, and communing with God his Maker." *

NOTE.—In addition to the authorities cited in the text and notes of the preceding lecture, the reader who desires a convenient summary of results concerning the Antiquity of Man, is referred to the following works:

L'Homme Fossile en Europe par M. Le Hon. This volume treats of the industrial occupations of the primitive Man, his customs, and his works of art.

Enthüllungen aus der Urgeschichte, oder: Existirt das Menschengeschlecht nur 6000 Jahre? von Dr. J. H. Thomassen, Leipzig, 1869. This book exhibits in a popular form the results of the latest scientific investigations upon the origin and development of Man. These are presented under a two-fold division—the Natural History of the Race, and the History of Civilization. All facts of recent discovery are methodically arranged with the proper authorities. The writer inclines to the notion of a great antiquity and to the theory of development. Yet he affirms that the derivation of Man from the ape must be purely a question of *probabilities,* and he disclaims Materialism, following Schaffhausen in these words: " The investigations of natural history do not concern themselves with the divine, the heavenly derivation of Man, but only with the earthly, the natural. And why should it be

---

* Professor Dana, Geology.

deemed unworthy of Man to regard him as the last and highest development of animal life? Did he come forth any the less good from the hand of the Creator, if in the dark womb of untold ages the animal type was more and more ennobled, until that human form was attained which Man regards as the image of his Maker?"

Die neuesten Forschungen und Theorieen auf dem Gebiete der Schöpfungsgeschichte von Dr. Friedrich Pfaff, Frankfurt, a M., 1868. After a careful statement of the discoveries bearing upon the Antiquity of Man, Dr. Pfaff infers that Man did not appear till after the ice period. He declares the uncertainty of all geological calculations intended to fix the period of Man's origin, and refutes Lyell's arbitrary estimates from the present rate of formation in drift and deltas. He finds no traces of Man, with any certainty, farther back than the great climatic changes of the Quaternary period, "the most reliable of which do not reach back more than 5,000 to 7,000 years from the present time."

The publication of Professor Arnold Guyot's lectures on *Man Primeval*, delivered in 1869 before the Union Theological Seminary, will furnish the public with the best exposition yet made of the discoveries of modern Science in their relations to Biblical history. Some of the most important suggestions on this point in the preceding lecture were derived from conversations with Professor Guyot, who is ever ready to dispense his knowledge for the advancement of Truth and Religion.

LECTURE VI.

The Sabbath Made for Man.

Gen. ii. 1. Thus the heavens and the earth were finished, and all the host of them.
2. And on the seventh day God ended his work which he had made; and he rested on the seventh day from all his work which he had made.
3. And God blessed the seventh day, and sanctified it: because that in it he had rested from all his work which God created and made.

WE now fairly enter upon the History of Man, to which hereafter this book is devoted—especially Man in his relations to the Kingdom of God. The Creator himself is here presented as resting from His work and instituting a commemorative Day of Rest—not for His own purposes, but in the interest and care of Humanity. The work of creation is now contemplated with specific reference to Man and his uses and obligations. "The heavens and the earth were finished, and all the host of them"—that is, the heavens and the earth as Man beholds them and is interested in them. We are not here to understand by the heavens the whole starry universe, as now made known to us by the calculations and appliances of astronomy, but simply the heavens as visible to the eye of Man, and related to Man and his habitation. These are sketched in outline before us as the work of Almighty God.

This work was now *finished*. There had been, as we have seen, a course of operations bringing the physical creation to a state of preparation for Man, its destined occupant and lord. At each successive stage in this work—after the dry land was separated from the seas—after the earth brought forth grass, herbs, and fruit trees—after the heavenly bodies appeared in their relations to this world, for signs and seasons, and for days and years—after the waters brought forth abundantly moving creatures, and fowl were created to fly above the earth—and again after the earth brought forth the living creature, cattle, and creeping things, and all manner of beasts—at each successive stage in this grand process God had pronounced it GOOD; and now, with the appearing of Man upon the highest platform of this physical creation, the whole work was declared finished. At any previous point in this process, to the view of an angel the world might have seemed incomplete and aimless; yet not wholly aimless, inasmuch as Man himself was prefigured, both in his physical structure, by the homologues in the animal creation, and also in the general order and arrangement of things for the support of such a being. At all events, we can now trace the preparatory steps in the adaptation and structure of the globe for the advent of Man.

The work was "finished" also as to the Divine plan and arrangement in respect to elements and materials. I have before noted the fact that Science has not yet ascertained that any new distinct species has appeared in the lower creation since the advent of Man; and although Science has developed to Man elements and materials in the constitution of the earth that to him were new, there is no evidence of the creation of any new elements or materials in organic nature since the appearance of Man. The whole work was finished

"All the *host* of them;"—this expression denotes both the splendor of the heavenly bodies and their orderly array. It is a figure derived from the marshaling of an army in which both these features appear—splendor and order. It is a very common figure of speech in the Bible applied to the heavenly bodies. "By the word of the Lord were the heavens made, and all the host of them by the breath of his mouth."* "He

---

* Psalm xxxiii 6.

telleth the number of the stars, He calleth them all by their names."* "Lift up your eyes on high, and behold who hath created these things, that bringeth out their host by number; He calleth them all by their names."† The Bible recognizes the order and unity of plan which are found in the physical universe. Thus David, in the 8th Psalm, sings, "When I consider Thy heavens, the work of Thy fingers, the moon and the stars which Thou hast ordained." The term "fingers" denotes carefulness of detail; the term "ordained," the established order in the heavenly bodies. As a shepherd boy upon the plains of Bethlehem, in the wakeful hours of night, David had gazed upon that wondrous sky which overhangs the land of Palestine, the transparent depth and pureness of the atmosphere, the number and splendor of the stars, the brightness of the moon, the glory of the constellations, the Pleiades chiming the advent of the spring, Orion girding himself in the autumnal skies as for battle with the storms, Arcturus guiding his sons in their nightly march around the pole; and these contemplations impressed him with the orderly arrangement of the heavens—the "work of the fingers of God."

The Creator is represented in the text as contemplating His work with infinite satisfaction. He did not withdraw, like Bramah, into a state of unconsciousness after having manifested His power in the production of the physical universe—for Bramah was but the *manifestation* of God in creation, while Vishnu is now being manifested in preservation, and Siva will be his manifestation in destruction. The God of the Bible is not an impassive being, indifferent to the works He has formed; but He rejoices in all the work of His hands. The sublime song of heaven recorded by John in the

---

* Psalm cxlvii. 4.   † Isaiah xl. 26.

Revelation, the song of the four living creatures, and the four and twenty elders is, "Thou art worthy, O Lord, to receive glory, and honor, and power,-for Thou hast created all things; and for Thy pleasure they are and were created." The use of the word "pleasure" here can not be explained as mere anthropomorphism. God is a being of soul; He has feeling, and He expresses it; and when He looks upon the works He has made, He is glad. His infinite heart dilates with joy in the contemplation of the beauty, the order, the majesty, and the splendor of the work of His hand.

On the seventh day God ended His *work* which He had made; and here began a new period, in the midst of which we now are. "He rested on the seventh day from all His work which He had made." To "rest" here does not mean to seek repose from fatigue, but to suspend activity in a particular mode of operation, to cease from doing thus and so. The Creator has not withdrawn himself from the supervision of the world and Man. As He is not indifferent to the beauty and order of His work, neither is He indifferent to the actions of His creatures. The Bible is full of the doctrine of God's continued providence over the creatures that He has made. In this sense, therefore, as the constant and universal preserver, He could not be said to rest. His tender mercies are over all His works. Our Saviour taught this doctrine of a constant personal providence, in His sermon on the mount, carrying this out even so minutely as to numbering the hairs of our head, and watching the fall of a sparrow; and it was in speaking of the Sabbath itself that He said, "My Father *worketh* hitherto, and I work." From the first moment of creation unto this hour God has continued His supervision of the works of His hand. The suspension of created activity, therefore, constitutes the "rest" spoken of in the text. The present course of things is the proper Rest-day of the

Almighty. He has put forth no creative energy since He brought Man into being; but at the end of the world, in the changes that shall produce a new heaven and a new earth, God will resume that creative activity which is now in suspense. Until then He rests. As one has said, "What we call a course of nature is the very Sabbath of God, nature itself being that holy pause in which God rests from His creative energies, that ineffable repose in which, though superintending and preserving, He *provides* for Man through law that he can comprehend, and an executing word that he can devoutly study."* "God rested on the seventh day," that is, as the last clause of the verse explains it, "He *ceased from all his work* that He had made."

"God blessed the seventh day, and sanctified it." Obviously this could not have been for Himself, but for man in relation to his Maker. To bless a day was to set it apart to be a blessing; but there was no sense in which God could make any one portion of duration more of a blessing to Himself than another, He being always self-contained and infinite in his blessings. To "hallow" the day was to dedicate it to some sacred, moral, and beneficial use; but of course God could not have made one period of time more holy than another to Himself. The sanctifying must have had reference to its use by and for others. This sacred day is God's day, which man should *devote* to Him in some special or uncommon way, turning aside from the common occupations of life to a separate peculiar observance of this portion of time. Hence this grand day of the Almighty, this on-going day, this day which, dating from the creation of Man as an intellectual creature, shall continue till the world and the present course of time shall close, is the type of the Sabbath, the Rest-day

---

* Prof. Tayler Lewis, in Lange's "Commentary on Genesis."

for the creatures of God. The blessing and the hallowing was the solemn establishing of the institution, since such a formality would hardly have been entered upon for a mere passing occasion. It was with reference to an institution to be continued through after-times; and the proof of this appears all along in the early history of the race. For instance, we trace the division of time into weeks, in the account of the flood, where Noah is said to have sent forth the dove at intervals of seven days. Again, in the life of Jacob, we find mention of a week as a recognized division of time, and so in other portions of the early history of the world. Some have supposed that this division was suggested by the phases of the moon, the lunar month being subdivided into four equal periods. But the phases of the moon, at the point of transition from one to another, are too obscure to have suggested this as a division of time so early in the history of the race. That would imply a knowledge of astronomy, which we can hardly suppose to have been then attained. The year and the month are marked off on the great dial of the firmament, as is the shortest division of the day and the night. The lights in heaven are for signs and for seasons, and for days and for years; but there is nothing in the phenomena of Nature which corresponds to the seventh-day division in a manner so striking as to have impressed upon an unscientific observer such a measurement of time. This would require much nicety of astronomical observation; and hence we must regard the week as an arbitrary division, and look for its explanation in some other quarter. The week was a wide-spread usage among the nations of antiquity. The Egyptians and the Hebrews had it, and so had other early people of the East. It was well known, also, far back in Hindoo and Chinese history. This general consent of antiquity to a division of time which is not strongly marked as a division of Nature, can best be accounted for upon the supposition of some common tradition as its source; and what more reasonable than the statement of the text, the designation by Jehovah of a sacred day to be observed by man from the beginning of the world? The exceptions to this seven-days period in the history of nations are just enough to prove the rule, for the usage prevailed among those nations that were connected most nearly by language and emigration with that part of Asia which was the cradle of the human family.

Again, the Fourth Commandment treats of the Sabbath as an institution already known. "REMEMBER the Sabbath day to keep it holy." This is not merely an emphasis for the future. It does not mean simply—keep in mind hereafter this day with a view to its sacred observance; the word remember recalls the past. An institution entirely new would have required a different phraseology. For instance, it would have been enjoined in some such language as this: Thou shalt keep a holy rest every seventh day. But the Sabbath day was recalled as an institution known to their fathers, and formerly to themselves, to be "remembered" as something that ought to be known, but had been allowed to slip out of mind. We find mention of this day in the history of the Israelites in the wilderness before the giving of the law at Sinai. When the manna appeared, it is recorded that "on the sixth day they gathered twice as much bread, two omers for one man, and all the rulers of the congregation came and told Moses." And he said unto them, "This is that which the Lord hath said, To-morrow is the rest of the holy Sabbath unto the Lord;" and on the following morning Moses said, "To-day is a Sabbath unto the Lord; to-day ye shall not find it in the field. Six days ye shall gather it; but on the seventh day, which is the Sabbath, in it there shall be none."* From this it is

---

* Exodus xvi. 22, 23, 25, 26.

evident that the institution of the Sabbath had long been known, although its observance may have pretty much died out among the children of Israel during their sojourn in Egypt. Now, it was revived with the memory of the patriarchal times and the history of creation, and reinforced by specific command as an institution to be remembered. Furthermore, the primary reason for keeping holy the Sabbath does not at all pertain to the Jewish commonwealth, but belongs to the history of Humanity. It existed from the day of the first Man, and is perpetual in its nature and obligation. "In six days the Lord made heaven and earth, and the seas, and all that in them is, and rested the seventh day; wherefore the Lord blessed the Sabbath day, and hallowed it." This was the primitive reason—it is still a prominent reason. Other reasons supplementary to this have from time to time been given for the observance of this day; for instance, the deliverance of the children of Israel from their bondage in Egypt; and since the Christian era, the resurrection of Christ; for the resurrection of Christ includes within itself, as antitype, both of the preceding grounds for the observance of this day; that resurrection symbolizes our deliverance from bondage, and our new creation into a higher spiritual life; and so grandly expressive is this symbol, that the day of the week has been changed to correspond with it. But the period of time to be observed as a Sabbath is altogether secondary; whether it be the seventh day or the first day is of minor importance. The essential point is the setting apart for sacred observances of a seventh portion of time; and the prime reason is as old as Man, and universal as Mankind. The day is primarily one of rest. It is agreed by scholars that the word "Sabbath" signifies etymologically "rest," and hence the day of rest, and by analogy rest from that which had commonly occupied the hands, the mind, the heart; and in harmony with this the

Fourth Commandment requires rest from the common labor of life. Now, whether we regard the Fourth Commandment in form as binding upon us as to its details or no, it matters not as to the essential spirit and obligation of this primeval institution; for rest is not accomplished simply by refraining from outward physical toil. The idea of rest pertains to the spirit; and the Sabbath does not become a real, substantial, refreshing rest, unless the mind rests from care—unless the spirit is released from trouble, anxiety, burdens, and toils. How welcome an institution that contemplates such an end! how beneficent in its adaptation to man! What a gracious invitation to every weary, burdened human spirit to enter upon a stated weekly period of rest! Hence it is also a day of rejoicing. A reason given for its observance at first is a joyous one—the fact of creation, the blessings of existence. Who has not known at times—especially when experiencing the full tide of health, in the bright rich calm of a summer day—the joy of simple existence? and since this is the foundation that underlies all faculty and possibility of enjoyment, it is well that we should be periodically reminded of our obligation to God for life. The contemplation of the glorious works of the Creator is fitted to awaken joy. Like the Psalmist, we should *consider* the heavens, the work of God's fingers. We should accustom ourselves to the contemplation of the Creator in the physical universe, as well as in the constitution of our own bodies and minds; and such a contemplation will ever be to the thoughtful and devout mind an occasion of rejoicing. As the week draws to a close, the laboring man begins to anticipate the Sabbath, and counts the hours that shall bring to him this welcome day of rest. What a mere drudge and slave would Man become if he had no such opportunity! It is impossible that the opportunity should exist except by some common understanding, some mutual arrangement pervading society; and this is provided for by Divine appointment at the first—that Man should have release from toil and enter into rest.

All right spiritual emotions and exercises have in them far more of joy than of constraint. To a mind rightly constituted, there is nothing burdensome in a spiritual religion. The true religious fervor is always joyous, and hence under whatever aspect we regard it, the Sabbath, as originally constituted and designed of God, is a day of welcome and blessing to Mankind. In every faculty and aptitude of his being we read that "the Sabbath was made for Man."

Human physiology and the laws of animal life teach us the beneficence of such a day. Some look upon the Fourth Commandment as restrictive and severe, but it is no more so than a good sanitary law or the prescription of a physician. If, for example, your physician should say to you, "Your constant toil is wearing upon you; your unremitting attention to business is affecting your brain; you must have rest; you must allow yourself so many hours every day for sleep; every now and then you must take a day or a week of relaxation; indeed, it may be wise for you to break off altogether for months of rest; otherwise, you will break down your nervous system; you will bring on softening of the brain, or paralysis and death;"—if a physician should lay such injunctions upon you, would you deem these arbitrary and severe? Nay, would not the timely caution, intended to regulate and save your life, be most beneficent for you? Now, the Sabbath anticipates these dangers and necessities for every Man, provides for him betimes the necessary and seasonable repose. It is emphatically in the interest of the workingman, and even also of the brute creation. "Man needs not only the periodic rest of sleep at short intervals, but longer rest at longer intervals; and, what is as important to health as sleep itself, he needs *change* of mood."* Nothing so preys upon the vital powers as a continuous strain of the faculties in one direction—to be forevermore upon the same thing, to be constantly working at the same point. The mind itself needs diversion as much as the body needs rest. The economic welfare of society is promoted by the Sabbath. Lord Macaulay finely said, "We, in England, are not poorer, but richer, because we have, through many ages, rested from our labor one day in seven. The day is not lost. While industry is suspended, while the plow lies in the furrow, while the exchange is silent, while no smoke ascends from the factory, a process is going on quite as important to the wealth of nations as any process which is performed on more busy days. Man, the machine of machines, the machine compared with which all the contrivances of the Watts and Arkwrights are worthless, is repairing and winding up so that he returns to his labors on the Monday with clearer intellect, with livelier spirits, with renewed corporeal vigor!"† The statistics of husbandry, of trade, of commerce, of every branch of business, will show that the profits are not diminished, but rather enhanced, by a due observance of this seventh period of rest.

And this beneficent design of the Sabbath is all the more marked, that rest from labor is *enjoined* upon moral grounds and for religious ends. The bulk of Mankind require a law to lead them to do what is wisest and best for themselves; for in such matters most men are in a condition corresponding to childhood; and just as—to recur to our former illustration—sanitary regulations must be enforced to teach the poor cleanliness, and preserve them from the visitation of epidemics, so for the most of Mankind this institution,

---

* See the fine article on the Sabbath in Dr. W. L. Alexander's edition of Kitto's Cyclopædia.
† Speech on the Ten Hours' Bill.

designed for their highest welfare as physical beings, and for their best mental and spiritual development, must be enjoined by authority in order that it may be duly observed. And the discipline of the soul through a government by law is of great importance in forming a strong and high-toned character. Every man who has known an inward conflict with temptation, the struggling of his soul against evil, has valued the strength of law, girding him about to sustain him; and if in the spirit of obedience to divine authority he has come out of such a struggle successfully, he is the stronger and healthier by reason of the discipline of law. Spiritual culture also thrives best by the help of special times of meditation and devotion,—as for instance the favoring hour of even-tide. The stated recurrence of such a season, the suspension of labor with reference to it, favors the improvement of the mind in its highest spiritual interests.

Such were the original grounds of the institution of the Sabbath; such are the reasons established in the very constitution of Man and of society for its proper observance. From all which we infer that *the sacred keeping of the Sabbath is due to the Dignity of Man, as a creature of intellectual and moral faculties.* The institution of the Sabbath dates from the beginning of rational existence upon the globe. Until Man appeared, there was no call for such a day, no faculty in any creature that could comprehend its meaning, no spirit upon the face of the earth that could fulfill its design; all things heretofore had moved under the ordinance of physical laws; but now, having created Man in his own image, God, as it were, *compliments* his new creature by lifting him above the plane of mere physical law—although in his lower animal nature he is still subservient to this—and addressing him as a being capable of spiritual communion with Himself, and destined to find his highest enjoyment and fulfill the highest plane of his being in fellowship with his Creator. For this purpose God set apart a day upon which He invites Man to special communion with Him, to the recognition of his spiritual dignity as a child of God. The Sabbath, as God designed it, so far from degrading him or imposing upon him a severe legal obligation, releases him from the sphere of mere animal life in the senses and the control of physical laws, and elevates him to the proud consciousness of his high origin and his immortal destiny. Man therefore best consults the dignity of his own nature by observing such a day in the spirit and according to the intent of his Creator.

I shall not here enter upon the question of civil legislation concerning the observance of the Sabbath. It may be stated, as a general rule, that we are not called upon to enforce divine laws as such by civil penalties, although the civil law against murder, and other criminal legislation, has a divine sanction. What I do insist upon is, that the divine law of the Sabbath, properly construed, is not irksome nor humiliating, nor a restriction of just liberty, but tends to enfranchise and ennoble the soul.

It should therefore be freely, voluntarily observed by men, not as an ordinance of human society, but as a gracious appointment of the benevolent Creator. The Sabbath calls for the devout and grateful recognition of God as the Creator of all things, and especially as the Father of our spirits. Nothing is more brutish in Man than to live with no acknowledgment of God. Brutes never lift themselves above their instincts; and if Man refuses to lift his soul to fellowship with his Creator as a spiritual being, what better is he, in this regard, than the brute beneath him? It is worse than brutish to live without gratitude. One can teach a dumb animal to recognize favors, to love and follow his master; and for Man, the creature of intelligence, the creature of affection, the creature of moral endowments, to feel no gratitude toward Him who has bestowed upon him all these faculties and capacities, and who so enriches and adorns his life, is to place himself below the level of the brute. Turning, therefore, from the animal, the sensuous, the physical, let us devoutly, thankfully lift ourselves into the sphere of spiritual affections, powers, aims, and hopes, by the worthy recognition and observance of the Sabbath Day.

LECTURE VII.

Woman and the Family.

Gen. i. 27. So God created man in his own image, in the image of God created he him; male and female created he them.

28. And God blessed them, and God said unto them, Be fruitful, and multiply, and replenish the earth, and subdue it: and have dominion over the fish of the sea, and over the fowl of the air, and over every living thing that moveth upon the earth.

ii. 18. And the Lord God said, *It is* not good that the man should be alone; I will make him a help-meet for him.

WE have seen abundantly, in former lectures, how the earth was designed to be *the abode of Man*, and shaped to that end. Glaciers, icebergs, floods, and fires had wrought their work upon the face of the globe; continents had risen or fallen with the changes of the seas, and other physical phenomena of which Geology finds the record had taken place upon the grandest scale; but this narrative, regarding these as of no account with reference to its great end, confines itself to Man and his development, in all his capacities and powers as a spiritual, intellectual, and social being. We have seen that he was introduced upon the stage of existence as a spiritual being, the express image of God, bearing His likeness in capacity, and intended to bear His likeness in character. His intellectual nature is at once exhibited and honored in his assigned dominion over the world, and in the faculty of speech. This last is brought into view in the close of the second chapter. It is perhaps idle to inquire whether Language was an immediate gift of God, or was originated by Man through the use of the faculties that God had prepared for this end. We are told that "out of the ground the Lord God formed every beast of the field, and every fowl of the air, and brought them unto Adam to see what he would call them." Every faculty of Man requires something objective to bring it into play. As light and the eye, the atmosphere and the ear, stand related to each other, so throughout the whole range of physical organization we find traces of adaptation to the faculties of Man. The adaptation of Man's organs to language might have remained forever inoperative had not objects been presented to him which called for such discrimination as could be given only by words. So all living creatures were brought before Man to be named; and whatsoever Adam called every living creature, that was the name thereof; and this discrimination among the cattle, and the fowl of the air, and the beasts of the field, expressed in words, was the rudimental beginning of language. This statement corresponds very well with the philosophy of language established by modern philology. When we consider how much is involved in human language, in its grammatical structure, in its application to the various uses of thought and of life, we find in this early use of language, a strong testimony to the intellectual nature of Man.

We have seen also, in the last lecture, that the religious wants of Man were provided for at the beginning by the institution of the Sabbath; that as a creature having moral needs and moral aspirations, he required some set time for intercourse with his Maker and spiritual meditation; and this day, set apart at the beginning for such uses, is a grand index of the religious nature of Man.

We come at length to the provision for Man's emotional and social nature. As his spiritual nature was recognized and developed by the presentation of God to his mind as an object of religious contemplation, and by the institution of the Sabbath for his religious culture; as the invention of language was called out by the presentation of objects to be named, so now his affections are to be brought into exercise by an

object worthy of them. When all was made ready, "God created Man in His own image, in the image of God created He him, male and female created He them." Thus far in the wide range of creation there had been nothing corresponding to the nature of Man himself. He could apply to the animal creation the terms of intelligent speech, but could receive from them no intelligent response. By degrees such animals as were domesticated would learn to respond to the names that Man applied to them; but beyond this they could have no comprehension of language. Of language as the vehicle of thought, of language as the means of intercommunication between minds, of language as the correspondence of spiritual natures, they could have no conception and no use, and so there was not in all the wide creation a creature having affinities of soul with Man. In the impressive words of the text, "for Adam there was not found a help-meet for him."*
This phrase is significant; a helper corresponding to him, his counterpart—not his double, not a mere repetition of himself, but the complement of his own being, corresponding with himself in all essential particulars, but at the same time supplying certain elements for social life and spiritual intercourse in which he himself was lacking.

By this double yet single creation, this duality in unity, as it has been styled, God set apart Man from all other creatures in the high and sacred institution of the family. There are sexes in animals and in some species of plants; a law of polarity pervades even in organic nature; but of no animal or plant are we told that God created but a single pair, and solemnly consecrated these to each other with His own command and blessing; and nowhere in the animal or vegetable kingdom do we find a uniform distribution into families; and when there is an approximation to this, it is purely instinct-

---

* Genesis ii. 22.

ive, without any rational or moral affection. But in the human species God created but a single pair; created the *two* that they might voluntarily become one;—one, not by a mere law of Nature, but one by a rational and moral choice in a union, not of chance, of convenience, or of impulse, but a union of deliberate volition and of sacred affection, and therefore indissoluble. Such was the nature of this union from the beginning. In a sense one might say Adam and Eve had no choice; yet was the union voluntary. Each saw in the other perfection, and this was the basis of their moral union. Adam had tasted solitude; he had given a name to every living creature; yet not one of them all had answered by intelgent recognition; there was not found a help-meet for him. But when the Lord God brought to him the woman, his heart at once said, "She is mine, bone of my bones, and flesh of my flesh;" he gave her a name derived from himself; and as the great prophet of his race, declared that "man should leave even father and mother, and cleave to his wife." The second Adam interprets this as the command of God, sacred from the beginning. And not only was the institution of marriage then ordained as sacred and indissoluble, but the reason assigned for this is of the highest moral nature,—that the image of God might be perpetuated in the world, the spiritual likeness following the natural from generation to generation. This is the argument of Malachi for the inviolability of marriage. Did not God at the beginning make but *one* wife for Man? Yet had he the residue, the fullness, the excellency of the spirit. He might have multiplied the human species without number by separate acts of creation; yet He made but one man and one woman. And wherefore, asks the prophet, did he make but one? "That he might seek a godly seed," * —that the family might become the educator

---

* Malachi ii. 15.

of the race in the knowledge and the love of God; that the father who should transmit to his son his own physical likeness, might also impress upon him the likeness of God; that thus holiness might be hereditary and perpetual; that the life of God, beginning in that mysterious and sacred union in Eden, might flow on in channels of love and purity till the end of time. Alas, that the fountain became corrupt, and has sent forth streams of bitterness and death!

God put immediate honor upon the relation He had so constituted. Beneath the feet of Man were countless generations of creatures buried in the rocks and soils of the Preadamic earth. These monster creations had passed away; Man could not have lived in their period; Man does not exist under their laws. They paved the ascent to his majestic throne. And now, mated with his own flesh, male and female created in the image of God, Man is inaugurated king over this earth that entombs all former creations, and over all its living things. A new era opens in the history of creation; an era that, beginning in the domestic constitution as its germ, shall evolve into the momentous history of the race in society, in the state, and in the kingdom of God.

Two facts of perpetual significance were established by the creation of Woman,—the fact of *sex*, and the fact of the *interdependence of the sexes*. One may conceive it possible for the Creator to have multiplied the race of Man by successive acts of creation; to have peopled the earth by some direct exercise of creative power with multiples of men simply; but He chose to create Woman as the medium for the increase of the race; and thus the distinction of Sex is permanently established in the race itself. And with this distinction comes in the condition and the feeling of interdependence,—neither sex complete without the other, neither sex possible to be continued without the other, neither sex able to represent the other, neither sex able to dispense with the other. No modification of society can ever displace these two fundamental facts—the sexes and their interdependence—and no society can endure which disregards or attempts to modify facts thus fundamental in the existence of the race itself.

The *status* of Woman hinges upon this vital fact of sex. Yet "the tendency of legislation in all modern states is to reduce marriage to an instrument for the legitimization of children simply, leaving all the relations of husband and wife which are not necessary to this end to be regulated by individual will. . . . In all European countries there is every day a stronger and stronger movement toward the liberation of Woman from all legal incidents of matrimony which are not necessary to prove the paternity of her children and provide for their maintenance." * But why seek to legitimize children? Why not let them belong to the state, as in Plato's Republic? Can they be cared for without love—that delicate, reciprocal love of the parents which results from their mutual fitness and dependence, and attracts them both to their offspring? To reduce marriage to a compact of legitimacy is to degrade Woman to a commercial creature, and rob her of that divine beauty which her sex has stamped upon her. No picture of a Man ever affects us as does the Madonna, crowned with the dignity and glory of Womanhood. To regard husband and wife as separate units held together by an artificial contract for purposes of legitimacy, is to destroy the true unit of Society, which is the Family. This is the fatal fallacy in Mr. Mill's essay,† that he overlooks those physiological diversities that both separate the sexes and provide for their union in a true equality.

---

* The *North American Review* for July, 1867. † "The Subjection of Woman."

"The foundation of this new constitution was laid," says Harris, "in the divinely instituted union of husband and wife. 'Have ye not read that He who made them at the beginning made them *a* male and *a* female [as intending to prevent both polygamy and divorce], and said [as the formal authentification of the great law of marriage already inserted in the constitution of human nature], for this cause [or, on account of entering into the married state] shall a man leave his father and mother [the nearest relation he had previously sustained], and cleave to his wife, and they twain shall be one flesh; wherefore they are no more twain, but one flesh.' A union this so intimate, that every other is to yield to it; so sacred, that the Divine Proclamation concerning it is, 'What God hath joined together, let no man put asunder;' so indissoluble, that nothing is to separate it but that which separates the soul from the body; so spiritual in its ultimate relation and aims, as to find its antitype only in that divine union which, as the fruit of redemption, is to survive every other, and to obtain its consummation in heaven."

The apostolic precept is, "Husbands, love your wives, even as Christ also loved the church;"* and the glorious church of the redeemed in heaven is "the Bride, the Lamb's wife."†

The Word of God always puts honor upon the institution of the Family. Both the moral and the civil code given to the Israelites in the wilderness guarded this sacred institution. Under the peculiar constitution of the Jewish nation, as a people separated from other nations, their origin, their religion, and their institutions, the interests of property, inheritance in a tribe, the distinction of nationality, pride of ancestry, and the hope of an illustrious posterity, combined to give honor and sacredness to the Family.

---

* Ephesians v. 25. † Revelation xxi. 9.

This relation, thus constituted by the act of God, rests clearly upon the *fitness of things*. It was not the mere arbitrary will of God that placed the marriage union of one pair at the foundation of human society, making this the channel of both physical and spiritual life to the race, but it was the wisdom and the love of God consulting in this, as in all the ordinances of creation, the perfect fitness of things and the best good of creatures.

Contemplating the Family as thus constituted, in its adaptations for happiness, the relation originates in love; a love whose strength and fervor lie in its individuality; whose very essence and beauty are that however multiplied in its examples, it is still in every instance unique and exclusive; that however intense and ample, its volume can flow in but one channel between two responsive hearts; a love which, like the subtile fluid of magnetism, may be conducted across the globe, over the prairies and the mountains, through the forests and the seas, and yet which finds its home and emits its spark only where the two poles of its being are brought together.

In this Family so constituted and so sustained, a two-fold want of the soul is met through *Society in Seclusion*. It is not good that the Man should be alone; and yet perpetual companionship with the busy world would merge his personality in the vast and complicated machinery of social life. To be himself, to keep the heart in play, to strengthen and develop the noble powers and sentiments of his soul, to keep the spiritual and immortal in a fine unison with the physical and temporal, Man *must* be, and yet must *not be, alone.*

There is no perfect happiness for an intelligent being where love is wanting. The pleasures of sense will not suffice the soul. These impart a real happiness only when the mind has etherealized from the grosser materials some spiritual essence,

THE FAMILY FOUNDED IN LOVE.

which serves as a bond or medium of intelligent and sympathetic intercourse with some other mind. Knowledge alone will not meet the soul's capacities and needs. The lore of ages may be poured into the mind only to congeal there like the glaciers of the Alps, whose crevasses and pinnacles reflect an awful lustre of emerald, but wear no living green. Only when the warmth of Love dissolves the icy intellect, and its treasures percolate through the sensibilities and affections of the soul, do the accumulations of knowledge diffuse a living freshness and joy. But to multitudes the heights of knowledge are inaccessible. They have no skill to climb, no time to make the effort, nor can they afford the services of a guide. It is theirs to till the valleys, to work the mines, to fell the forests, to dredge the marshes, to bridge the rivers, to grapple hand to hand with the hard, rough world, to subdue nature while others enjoy it, to work at handicrafts while others reason of philosophy and discourse of art. And where shall these, the multitude of Mankind, find happiness, if knowledge, the pursuit of learning and science, be its only or principal medium? Sensual pleasure will but debase them and bring out more strongly the coarser features of their character. Sensual pleasure will carry them downward toward the level of mere brute machines. Where, then, shall these, the multitude of mankind, consigned to toil, cut off from the resources of knowledge, imbruted by sensual pleasure, where shall they find true happiness? Affection gives the answer. The circle of home opens to receive them, then closes about each in his proper sphere its golden bands of love. *Home*, that briefest word of our good old Saxon tongue—there lies in it the wealth of all language, of all affection, of all virtuous joy, of all pure memories, of all innocent hopes; the prattle of the infant, the gleeful laugh of childhood, the song of the maiden, the cheerful labor, the merry pastime, the sweet repose of evening when toil is ended, the united meal, the household stories, music, and diversions, the various ages, interests, and plans revolving about one centre, and that centre *love*. Whose eye does not moisten with unbidden tears at the thought of Home? These four letters are the chord of human happiness for every gamut; whenever the scale of life begins, these letters are its perfect consonance.

"God setteth the solitary in families." This was His institution at the beginning, and He ordained it to be perpetual. It is not good, said Jehovah, that the Man should be alone, wherefore He made a help-meet for him; *a* male and *a* female made He them. And this numerical relation of the sexes, which no science can control, maintains its balance in successive generations, through all the incidents of climate, of migration, of war; so that taking the world as a whole, you hardly find in adult age more of either sex than will suffice to constitute separate and independent families after the model of the first. The law of human life, and the appointments of Providence in its myriad incidents, concur in giving to the original institution of marriage a perpetual sanction. The poet represents this wondrous adaptation of two souls in the marriage tie as " perfect music set to noble words."

> " Self-reverent each, and reverencing each;
> Distinct in individualities,
> But like each other even as those who love."

The relation which is thus clearly founded in divine revelation, and in those physiological laws and those appointments of Providence that constitute the fitness of things, has the best possible adaptations for the education of mankind. Wherefore the prolonged dependence of the human infant? Why is he whom God has constituted the lord of creation kept for years in a condition of helpless infancy, while a few days or months suffice to bring to maturity the offspring of inferior animals? *Why* is this? Because this human infant is not to be a mere creature of instincts, but a being of reason, of affection, of will, to act under moral law with responsibility to Man and to God, and with relations to the Unseen, the Infinite, and the Eternal. The littleness of Man in his infancy points to the grandeur of his future. For that greatness he must be educated. Principles do not come by instinct, and instinct should never control the will. Reason and the will should be educated by the guiding hand of affection. All that the brute needs of a parent is to be fed till its animal strength and instincts shall enable it to get its own food. But Man is to move upon a higher stage. He belongs to a race that has a history. His birth is not merely a numerical addition to the race—it is a continuation of its history. And as history links him with the Past, so destiny links him to the Future. His life is not an isolated unit. He stands in the present to carry forward the lessons of the Past into the destinies of the Future. One can not leap full-grown to such a stature. He can not rise to it by instinct. He must be educated for it, disciplined to its high argument. His very dependence is a part of that discipline; in reverence, in obedience, in submission, in trust, in love, in self-government, in knowledge, and in virtue. Now, for this education the Family is provided, and only in the Family rightly constituted can this be fully attained. Dependent infancy, sympathetic and imitative childhood, inquiring and confiding youth, these demand the tender, loving, watchful, patient, and controlling nurture of the Family. Maternal love—that most mysterious and most potent of the forces that guide and control our being —is the only power upon earth that can fitly educate Man. God has given to the mother instincts and affections equal to that responsibility; and by thus matching the utter weakness of Man in infancy with the correlative strength of Woman's love, He again sets his seal to the primeval law of marriage. For outside of the marriage relation even the instincts of maternal love are stifled. Only when consecrated to virtue do these instincts become the educators of the race.

In every aspect, physical, economical, intellectual, moral, the Family is the appointed nursery of the race for the well-being of the individual and the progress of society. Neatness, politeness, industry, economy, order, punctuality, affectionateness, truth, where can these be acquired from tutors or from books, as they are instilled with the dews of daily affection in the household and warmed with the sunshine of its love? As one has said, "Home-education is a law of Nature. And where can that labor of love be found more minutely and wisely divided than between the father and the mother, between patience and power, tenderness and authority, the instinctive love of offspring, and the moral regard for the excellence and well-being of that offspring."

Hence it is that the Family, as originally constituted in Paradise, is the only true basis of society. No theory of the *social compact* has ever been devised that would stand the test of reason, of history, or of natural law. The whole constitution of the world is against such a theory. By that theory men existed as isolated units, till their common wants or fears, or the aggressions of the strong upon the weak, brought them together in compact communities. But trace the stream of history back through any of its channels, and you find society always emerging from the Family. The first birth into the world was a birth into society, which then began in the bosom of the Family. Since the creation of Adam, Man has never existed in the world as an isolated unit. It is a law of his being that he shall enter upon his existence in the social state, and there abide.

In the Family, Man is trained to reverence for just and lawful authority; submission to government; obedience to law. In the Family, one learns to control instinct and passion by reason and affection. In the Family, one learns to respect the rights and the feelings of others as equal to his own. The child in the nursery is taught the rights of property, and the claims of gratitude and love. Hence the Family has been aptly styled "a rehearsal for society; where inferiors and superiors of every kind and degree mingle and co-operate— youth and age, weakness and strength, ignorance and knowledge, male and female, affection and authority, are blended together into one compact society. . . . Every coming relation of life, every future form of duty, and every subsequent social affection are seen virtually put in rehearsal for the more public scenes of after life."

A favorite theory of what is known in Europe under the names of Red Republicanism, Socialism, and the like, is that society is only an aggregation of individual men, and that its laws and customs must be determined by the wishes of the majority of these individual units concerning their supposed interests and rights. But this theory provides no bond of coherence or continuity such as is required to constitute *society;* its union is a rope of sand; self-interest alone keeps these individual atoms together, and there is as much of repulsion as of cohesion in mere self-interest. What sort of society these units of individual men make when brought together by some motive of self-interest is seen in every newly-opened gold region or mining district, where men congregate in herds, but without families.

There is not in any society such a thing to be found as this individual unit. No man exists as a unit. Every man comes into the bosom of the Family as his first step into life, and there is put at school for membership in the wider family of the state. Hence, if we would have a Community presided over by justice, maintaining equal laws, guarding the individual in the many, and all as one, then must we maintain intact and sacred as it was in Eden the institution of the *Family*. Lycurgus sought to raise a warlike race by taking the infant from its parents to be reared in the gymnasium as a child of the state. Plato, in his Republic, advocates a system of Free Love, abolishing the Family, destroying weakly children, and taking the strong into a public nursery, where they may be reared without natural affection. A society of brutes and bullies, a government of gladiators and tyrants, would spring full-armed from the bloody ashes of the Family.

The Family is the proper nursery of the race in morality and religion. For this it was designed by the Creator. Here are inculcated those principles of moral government; here are developed those pure and generous affections; here are nurtured those immortal hopes, that fit the growing mind to recognize and assume relations toward that heavenly Father of whose authority and benignity the earthly father is its daily type. The Family was instituted as a school for heaven, whose perfect symbol is the Family gathered in their father's house.

The study of Woman's primeval relation to Man and the Family would aid in the solution of some questions concerning her sphere in modern society. In discussing these questions, some writers overlook entirely the fact of Sex, which we have already shown to be fundamental, not simply as a physical distinction, but in its social and moral bearings. The advocates of the theory of the individual unit of society would have no longer women, but what Count Gasparin has aptly styled "*female* men;" in the struggle for technical equality all the finer distinctions being effaced, and only that remaining which Nature has indelibly stamped in the physical constitution. But the true interest of society demands that we should assure to Woman *that prerogative of honor in domestic and social life* which we have in part gained for her by redeeming her from a life of drudgery.* While labor is Man's primordial necessity, " *Woman's* right to labor " is a cry full of evil omen. It marks the deterioration of that manly sentiment which has hitherto accorded to Woman in this Republic a position of honor and prerogative unknown in the titled society of the old world. There she has the " right to *labor* "—as shop-keeper, stall-tender, street-cleaner in the cities and towns of France, and as a peasant in the fields; there she may labor at the oar upon the canals of Holland; there she may have undisputed right to labor over the vast plains of Germany and the steppes of Russia, digging, hoeing, ditching, and following the plow; there she has the scavenger's right to labor, in Switzerland, Egypt, Syria, gathering with her hands the ordure of animals for tillage or fuel; there, in Spain and Italy, she has the right to trudge weary miles after the cattle as they browse in the scorching heat or the pelting storm, or to burden her head with loads of wood or grain fit for the back of a camel. It shames me that, in this free Republic, where the sanctity of Womanhood has been guarded with a jealousy that the age of chivalry never knew, we are beginning to look upon Woman as a creature doomed to labor. Her "right to labor" is wide as the world, if she covet that. Let her go forth to labor, if she will, and produce hands and feet and features of corresponding coarseness; but in quitting the gentle occupations of the household, that she may compete with Man in every form of labor, she may assert a muscular right, with which she is but imperfectly endowed, at cost of a spiritual prerogative which is

---

* The author reproduces here a section of his oration before the Phi Beta Kappa Society in Yale College, published in the *New Englander* for January, 1869.

hers by Nature and by the concession of all noble men. That daily toil for daily bread which is Man's inheritance through the fall, was not laid upon Woman at the first; and it is no social enfranchisement, but a hardship imposed by a false condition of society, that would put it upon her now. Let Woman use her finer faculties in education, art, science, manners, the humanities; let her win here the place of preferment; and when she must perform manual labor for subsistence, let her be encouraged, respected, and remunerated in this also, as one bravely meeting a hard lot; but let us not dignify with the name of "right" a physical necessity that marks an abnormal condition of society. Sir Samuel Baker informs us that in Latooka " women are so far appreciated as they are valuable animals. They grind the corn, fetch the water, gather firewood, cement the floors, cook the food, and propagate the race; but there is no such thing as love." Shall we go back upon our civilization, back upon our Christianity, to the White Nile theory of Woman's labor? Such would be the result to Woman of that theory of "rights" which makes her equality with Man a reason for her "doing everything that Man now does."

The equality of the sexes is not *sameness* of endowments and adaptations, but *equality* with *differentia*. The attributes of sex belong to the soul as well as to the body, so that in their intellectual and spiritual natures, much as they possess in common, the Man and the Woman are also the complement each of the other; and in the distribution of these complementary qualities Woman certainly has no cause to envy her partner. Her delicate and beautiful presence, her graces and charms of person and manner, her intuitive affinities for the true, the pure, and the good, her divine faculty of counsel, her all-pervading, all-controlling influence—these are *prerogatives* which Woman has no right to vacate by reducing herself to a mere tool of productive industry, a numerical factor of political economy. Physiology demonstrates that Woman is not so constituted as to compete with Man in labor, since there is an appreciable difference between the two sexes in the proportion of red-blood-corpuscles, upon which depend both "*vital activity* and the capacity for *sustained exertion*," whether of muscle or of the brain.* But though Woman is thus inferior to Man in native vital force, a kindly Nature has imparted to her a more subtile vivacity and grace, showing that hers are the beautiful ministries of life, and Man's its rugged toil; and it is this prerogative of Womanhood that she would sacrifice by attempting the unequal strife and burden of the " working-day world."

Only at the cost of this same prerogative—the prerogative of ruling in society through the homage of valor to grace, of strength to refinement, of muscle to heart—only by sacrificing this could Woman enter into the arena of political strife. The delicate laws of her physical organization, the more subtile and beautiful laws of her social and moral influence alike forbid this uncrowning of her Womanhood. One who would claim the right of political action must be equal to serving the state in its demands as a civil organization. Here, emphatically, rights and duties must be correlative. Since suffrage carries not simply the act of voting but the function of ruling as well—not only declaring one's preference in political affairs, but actually governing the whole community—this can not be the natural right of any individual, but is a privilege to be accorded by society—by the Body Politic finding itself in power—in view of one's competence to serve the state in its

---

* See Carpenter's "Principles of Human Physiology," sixth London edition, pages 168 and 198. "The *maxima* in the female do not pass much higher than the *mean* of the male, while her *minima* fall far below his; on the other hand, the *maxima* of the male rise far higher than those of the female, while his *minima* scarcely descend below her *mean*."

rightful requirements, and with a wise and impartial consideration of the needs and welfare of the entire commonwealth. To enter political life argues capacity for civil duty; capacity to serve the state in the jury-box, in the police, in the camp, in the battle-field, in port-surveys and defenses, in the revenue service, in a routine of official duties that suffer no intermission; and Woman can not do this, can not trust herself to undertake the service for which she is physically incapacitated, can not be trusted with it with safety to the commonwealth. Witness, for instance, the protracted and exhausting session of the Senate upon the impeachment of the President! If she would fulfill the sacred functions of her nature, she can not accept the responsibilities of the public service, for the divine laws of physiology, and the divine constitution of the family, as the perpetual source of human society, can never be set aside. Either the vast majority of women must become wives and mothers, or society and the state must cease to be. But while Woman shall continue to fulfill for society that most serviceable, most honorable, and most sacred office of Maternity, which is hers by divine right, her very nature must forbid her employment in the public service of the state.

Reverting for a moment to the thought that, in this country, to vote is to participate directly in the *power of governing*, I maintain that the right to vote must rest upon ability to discharge the duties of citizenship in the service of society as a civil organization. This is the only logical foundation upon which the right of suffrage can be based. To base it upon taxation is to narrow all the great concerns of society down to the one point of mercenary interest. One may receive the full value of his taxes in public order and security, without being entitled to vote by reason of his assessment; and, on the other hand, taxes may be most unjust and oppressive, and the public order and safety most lax, where everybody votes, as in New York, and the representatives of the majority levy upon the property of the minority for their own schemes of plunder. To base the right to vote upon the abstract equality of individuals is to confound natural and personal rights with political powers; but voting is a *power* in the state which no one can inherit by nature. If Man is endowed by Nature with the right to vote, if this is a right that inheres in humanity as such, then by what authority can minors and paupers be excluded from the polls, or a term of naturalization or a degree of education be required for admission to suffrage? In the last analysis, the Political Society must determine for itself in whom this power of control over public affairs shall be vested.

Is it asked whence has Society this right? The answer is simply that the Body Politic which possesses the power to rule, *must* rule upon conditions of its own making; it is bound to make these conditions just and fair in view of all the interests of society, but the remedy for injustice can not be found in admitting everybody indiscriminately to the function of ruling as a "natural right." The same power in society which regulates suffrage in the case of minors, paupers, and others, can attach to suffrage such conditions and limitations as the general good may require. Hence the natural equality of the sexes has no bearing upon the question of suffrage, which rests on constitutional and other qualifications for the service of society as a civil organization. That the capacity for such service is denied to Woman is not a fiction of civil law but a fact of physiology, which no legislation can ever change. The notion that the equality of the sexes requires the equal distribution and exercise of all civil, social, and personal functions and rights, leads to absurdities the most grotesque and revolting.

But we are now concerned, not so much with the abstract question of Woman's entering into public life, as with the influence of this upon the tone of society in a republican government. The tone of national life, the very continuance of the nation, depends upon the position of Woman more than upon any other single fact; and it has happened to Woman thus far in the constitution of American society, to be a conservative, elevating, purifying power, by virtue of the prerogative accorded her of ruling by character and influence apart from the contest of numbers. In a country which has no traditions of feudalism and no forms of society nor government to inspire sentiments of veneration and loyalty, the spirit of chivalry—"that nurse of manly sentiment and heroic enterprise"—has found its expression in loyalty to Woman. This sentiment, ennobling and refining a democratic people, is of more value to the Republic than all the balances of the constitution. It belongs to the divine harmony of society; for the Creator has intrusted Woman to the honor of Man in the family and the state, for the culture of the stronger through care and consideration for the weaker. Man looks up to Woman with the homage that chivalry renders to the delicate, the beautiful, the spiritual, the true.

But if Woman, disdaining her loyal defender, shall enter the lists to contend with Man by sheer force of numbers, clamoring for rights, he will say to her, "Stand upon your own strength and fight your own battles, expecting neither loyalty nor chivalry from me." An editor, distinguished as much for his courtesy as for his generous sympathy with all enlightened reforms, was besought by a champion of Woman's voting to advocate her cause. To her repeated demands of "right," he replied with quiet and cogent argument; but with such pertinacity did she pursue him that he said to her at last, "Madam, I fear if you come to me again in this manner, I shall be compelled to answer you as if you were a *Man!*" That saved him further intrusion, and opened her eyes to the possible future of Woman, should she gain the right of being talked to like a Man! Sad would be the social state in which men would feel challenged by the position of Woman to deal with her on public questions as they deal with one another.

Even if the ballot could raise Woman politically, the nation can not afford so to degrade its men by divesting them of the sentiments of delicacy, of honor, of loyalty—in a word, of chivalry, and arraying the sexes in the contest of numbers. Woman can not hope to act for herself in public life, and still receive the honorable consideration now accorded to the delicacy of her sex. She must choose between the two; and if she shall elect the latter, she will inevitably find that in what direction soever she forces herself outside the sphere of delicate and chivalrous regard into the contention of labors and of numbers, she is taking a step toward her own degradation. If she can brave the opprobrium, society can not risk the consequences.

It is assumed that Woman will bring to the polls a soothing element and improve the moral results of elections. On the contrary, her greater intensity of feeling for *persons* would bring a keener acrimony into our political campaigns. We can not forget how the Women of the South incited the rebellion and inflamed its hatred and atrocity; nor that Woman produced the worst monstrosities of the French revolution; nor can we shut our eyes to the fact that in great cities the *Bridgets* would roll up the majorities of the demagogues, and that Washington would have its Maintenons and Pompadours to add their intrigues to its political corruptions. The history of church elections in which Abbesses had a voice, is a warning here. But the calamity to be shunned is that Men, ceasing to respect and honor Women in their prerogative of influence, shall fear or court them as an element of numerical power!—for when the spirit of chivalry with its generous loyalty to sex is gone, the glory of the Republic will be extinguished forever.

The elevation of the Poor requires that the institution of the Family be established and maintained among them in its original sanctity. The problem of pauperism in great cities weighs more and more upon the reflective and philanthropic. Many are the expedients of benevolence for relieving the poor without adding to their degradation, yet it is well-nigh impossible to administer charitable relief for any considerable period of time without in the very act degrading the recipient. As yet many of the plans of benevolence have been only expedients; they do not reach the seat of pauperism in society, and therefore accomplish little for its cure. A perfect remedy may not be possible; but certain it is that alms-giving, whether occasional or systematic, and the whole operation of charity, touches only upon the surface of this monstrous evil, and, as before hinted, tends to perpetuate if not to aggravate that which it seeks for the moment to relieve. Schools of industry begin at a deeper stratum than mere charitable relief, and working upward from this lower level they tend to elevate those who come under their influence. But these do not go down to the very foundation of the evil with which they attempt to cope. Asylums and other institutions for training children away from their parents are but expedients to lessen the evil in the next generation. They do not exterminate its roots. Religious visitation, tract and Bible distribution, reach here and there an individual, but how little impression do these make upon the enormous mass of physical misery! Nay, it is almost hopeless to remedy physical misery by moral means, or to elevate the moral feeling and condition of those who are perpetually dragged down by physical necessities. What hope is there of the conversion of one amid such impure and loathsome surroundings? Bring some tenant of a cellar or garret, reeking with filth, drunkenness, and loathsomeness, into the house of God; let his mind be aroused to some idea of his moral nature and religious needs; kindle within him a desire for a higher life,—then send him back to kennel for the week with all that wretchedness and infamy, and what hope is there of permanent good? How soon will the moral light, so faintly kindled within him, be quenched again in physical wretchedness? We must resuscitate the Family; we must encourage the poor to form their independent homes in cleanliness and privacy; we must erect suitable buildings for their accommodation at moderate rents; we must supplement these with the lodging-house, the wash-house, and other institutions that will answer the purposes of a home for the lowest stages of humanity; in one word, we must keep up at every point the associations and claims of the FAMILY, before we can hope for a radical change in society. The doctrines of Socialism do not meet the case; there is enough of community now, far too much of it, among the degraded poor. We need the restoring, the elevating power of family sanctity as at the first. To this end Christian philanthropy must direct itself, or with the growth of wealth and luxury on the one hand, we shall see upon the other a corresponding degradation of the poor.

How wonderful the love and care of God for Man as manifested in the first provisions made for the blessedness and sanctity of the human race! This Biblical narrative presents to us something higher and better than physical laws. It is by no means unmindful of the laws of Nature. It sets before us in sublime array the course of Nature, with a comprehensive brevity that allows for and may include all known physical laws and forces in the wide extent of their operation. Whatever has been brought to light by scientific investigation in the records of the globe is amply provided for in the terms and conditions of this narrative. The successive acts of creation here described may all have gone forward by that uniformity of procedure which we characterize by the name Law, which, after all, is but one mode of describing certain actions or effects of the Divine will.

This narrative, therefore, gives to us all that Science gives,—but more. Rising above the plane of physical agencies, which are throughout implied, it directs us, as we have seen, to the First Great Cause, the intelligent, spiritual Cause, and thus brings us to the highest point of metaphysical conception, the point toward which all investigation in the sphere of physical Science at length directs us, but which physical Science, as such, can not definitely determine. Here, where metaphysics has essayed its sublimest attempts at discovery, this narrative makes to us the simple revelation of the *one personal God*. But it does not rest here. Metaphysics may conduct to the conception of a spiritual Being, absolute in His intelligence, infinite in His nature, almighty in His power. The narrative gives us this, and more; for as soon as it introduces this Creator in relation to Man, it presents Him in the aspect of a Father. All through the course of the creation, at intervals it had pictured the Creator himself as delighting in His works, as having a benevolent joy in the perfection of that which He had brought into being. He looked upon that which He had made, and saw that it was good. But in Man the narrative presents the Creator as the Father producing a child in His own image, after His likeness; and then with a Father's thoughtful care, providing for every want of his compound nature, for his physical comfort and enjoyment, for the gratification of his tastes and his sense of beauty; for his affections and his social nature through the medium of the Family, and of that society which must grow out of the Family; and for his spiritual communing with Himself, the Father of spirits, in that beatific intercourse which was the privilege and joy of Man at the beginning.

In all these arrangements, designed to be perpetual, we behold the love and the care of God. Wherever these arrangements have failed of their beneficent purposes, it has been solely through the perversity of Man; and just so far as Man shall return to the original design of the Creator, in the institutions of the Family and of the Sabbath, in the maintenance of pure domestic love and pure spiritual worship, will human society be advanced in integrity and blessedness, and once more approximate to that Paradise which was the glory of its beginning, the tradition of which fills the broad pages of history, and the realization of which in the hereafter will be that golden age toward which all poetry and prophecy direct our hopes.

www.ingramcontent.com/pod-product-compliance
Lightning Source LLC
Chambersburg PA
CBHW030345170426
43202CB00010B/1248